The Meaning of Birds

Simon Barnes

The Meaning of Birds

PEGASUS BOOKS

NEW YORK LONDON

THE MEANING OF BIRDS

Pegasus Books Ltd
148 West 37th Street, 13th Floor
New York, NY 10018

The images in this book have been taken from the following sources:
An Illustrated Manual of British Birds by Howard Saunders (Gurney and Jackson, 1889);
The Natural History of Birds by Thomas Rymer Jones (Society for Promoting Christian
Knowledge, 1867); Wood's Natural History of Birds by J. G. Wood (Routledge, Warne and
Routledge, 1862); Wikimedia Commons (pp. 196, 261 and 320).

First Pegasus Books hardcover edition January 2018

Book designed by Ken Wilson | point918

ISBN: 978-1-68177-626-2

10 9 8 7 6 5 4 3 2 1

Printed in the United States of America
Distributed by W. W. Norton & Company, Inc.

Contents

1

Taking wing

Flight has immense meaning for us humans because we can't do it. Instead we live in a dream of flight, and flight envy is part of the human condition. That's why birds, more than any other group of living things, draw us into the world beyond humanity.

Go to the church of the Holy Trinity at Blythburgh in Suffolk. It stands on a small eminence surrounded by marshland. The estuary of the River Blyth stretches out before it, underneath the vast East Anglian sky. Inevitably it's a fabulous place for birds. You can birdwatch in the churchyard, picking up the flamboyant avocets and the stately, cruising marsh harriers. You can hear the sharp calls of redshank, curlew and oystercatcher, so that it sounds as if the marsh itself is singing to you.

Then step inside the church and raise your eyes to heaven; there you will see a flight of twelve angels soaring above you. Angels are the holiest beings in Christianity: second only to God in their sanctity. So what do we do to make them special? We give them wings.

Now look more closely at the wings that cleave to the lofty ceiling of the church and you will see that these angels are flying with the wings of marsh harriers. Across the centuries marsh harriers have lived around Blythburgh. They are birds of prey, and they are capable of flying at an uncannily low speed. They hunt by quartering the waterways and reed beds for water voles, moorhens and frogs, and their technique is to do so very, very slowly, inspecting the ground below them with meticulous attention. Flying slowly without falling out of the sky is one of the hardest

previous page:
THE SWIFT.
For caption, see page 260.

tricks a flying animal or a flying machine can pull off, but marsh harriers are masters. They are improbably buoyant. They fly as if they are not really at one with the earth beneath. Occasionally, and with great reluctance, they come down with the action of a shuttlecock in slow motion. They touch the base earth as if they weren't quite sure of their welcome. The same earth-rejecting airiness can be found outside on the marsh and inside on the roof of the church.

And in both cases the power of flight touches the human soul.

All human mythologies are filled with flying creatures. The stories we use to explain our world and our place within it are packed with flight. Every time humans sought to express something special, magical and full of meaning, the idea grew wings. Every culture on earth brings us creatures that waged war on gravity and won: Chinese dragons can fly without the need for wings, and they bring good luck to all those worthy of it. The thunderbird of the indigenous people of North America is not a puppet but a great bird whose wingbeats make the sound of thunder. Garuda flies through Hindu and Buddhist mythologies. Quetzalcoatl is the feathered serpent of Central and South America. The rainbow serpent—a snake that arcs across the sky—is a major figure in the mythology of Australian Aboriginal people.

The West is as full of flying myths as any other part of the world: fairies, our own dragons, harpies, hippogriffs, witches, the phoenix, the sphinx and Pegasus. We still make myths around the same theme: Superman is defined by his ability to fly; Harry Potter was the best flier at Hogwarts: Narnia acquired its magical protecting apple tree with the help of a boy: a girl and a flying horse called Fledge.

The West's essential flying story is that of Daedalus, who made wings and flew, and so was able to escape from King Minos of Crete. His name was borrowed by James Joyce for his fictional self, Stephen Dedalus, who flew by the nets that were set out to trap him. But Daedalus's son Icarus is the one we remember: the boy who flew too close to the sun. The wax that held his wings together melted and down he came: a tiny splash in the painting traditionally attributed to Pieter Bruegel the Elder, *Landscape with the Fall of Icarus*. It was commemorated by the poet William Carlos Williams as:

> *a splash quite unnoticed*
> *this was*
> *Icarus drowning*

The Icarus legend is about the eternal collision of human limitations and human ambitions

The Icarus legend is about the eternal collision of human limitations and human ambitions. The flight ambition is expressed in most of the games we play. All the non-confrontational sports are about flight, or at least the defiance of gravity. With long jump, high jump, pole vault and gymnastics we seek to fly, however briefly, under our own power. Winter sports give us the illusion of flight in the freedom from friction as we slip, slither and slide with skis and skates and sleds. In the horsey sports we borrow wings from another being: for every rider, every horse is Pegasus.

Flight isn't just a quick way of getting from A to B or a more efficient way of getting the groceries. It's an escape from the human condition, an abandonment of the cares of daily life: an entry into the world of good dreams in which you float effortlessly across oceans. Freud was mad about flying dreams:

> Do not feel disturbed because the dreams of flying, often so beautiful, and which we all have had, must be interpreted as dreams of general sexual excitement, as erection dreams… It is no objection to this conclusion that women may have the same dreams of flying.

Many drugs are associated with flight. Try Timothy Leary and *The Politics of Ecstasy*, Aldous Huxley and *The Doors of Perception*, *The Teachings of Don Juan: A Yaqui Way of Knowledge* by Carlos Castaneda: all books that affirm the desirability of mental flight. They seek to break down the barrier between objective and subjective realities, so that the question of whether a person 'really' flew becomes not only irrelevant but also meaningless. Like all the other ways of seeking flight, hallucinogenic drugs are about the search for an uncommon freedom. Mike Heron of the Incredible String Band summed this up in his 1966 song Footsteps of the Heron, in which he stresses the uniportance of money, apart from the fact that the stuff enables him to 'buy wings to fly, 'cause without them the sun ain't so sunny.'

Sex, religion, sport, spiritual exaltation, drugs, wish-fulfilment dreams, hope, ambition: all these things are tied up with flight. We took the idea from the birds, and always the idea leads us back to the birds.

facing page:
THE PEREGRINE
FALCON
is the world's fastest
bird; its diving speed
has been measured
at 186 mph.

Most ground-dwelling mammals have roughly the same vocabulary of movement. Horses show this most clearly: a walk has four beats, a trot has two, a canter has three beats and a gallop brings us back to four again. But there are all kinds of combinations and blurrings of the boundaries, as you see when you watch horses messing about in a field.

The vocabulary for flight looks pretty limited on the face of it. Roughly speaking, there are four ways to fly: powered flight, like a passenger jet or a blackbird barrelling away from you with a rattling alarm call; gliding with still wings, like a glider or a gull crossing a beach; soaring—that is to say, gaining height with still wings—which can be done by a glider and rather more efficiently by a vulture; and hovering, like a helicopter or a kestrel.

But there are many ways in which these classifications blur into each other and are used in different ways or in combination. So let's look at some of the champions of flight.

The fastest bird in the world is the peregrine. This is a falcon that has taken to the city centres of Britain in recent years. They can be found nesting on the spire at Norwich Cathedral and cruising down the Thames with the dome of St Paul's as a backdrop. They are birds that feed on other birds; and birds—since they can fly—are by definition a difficult target. Overcoming that difficulty is the peregrine's speciality, and speed is his solution.

A peregrine's flat-out speed in straight and level flight is comparatively modest, around 40 mph (64 kph). Where they excel is in the stoop. This is the move in which they assume the shape of an anchor or the Greek letter *psi* and make a crash-dive at a target some way below. They have been measured at 186 mph (304 kph) in such circumstances, and greater speeds have been claimed: there is a record of 242 mph (390 kph). Slow-motion footage shows that the peregrine continues power-flapping in the stoop to build up yet more speed. Even at the extreme end of its dive it has control over direction. When they make successful contact with the target bird, the shock of the impact will usually kill it outright.

It's not easy to deal with such speeds, as humans discovered when we started to make jet planes. The first problem comes from the need to handle a wind of 200-odd mph (300-plus kph). It's difficult to breathe at all, because at high speed a wall of still air builds up in front of the moving object. The designers of early jet planes found that at a certain speed the jet could no longer take in the air it needed for power.

Peregrines have a small projecting cone in front of their nostrils to break up the wall of still air, and that allows them to breathe. This leaves them with the problem of dealing with a 200-mph gale inside their own airways. They solve this by means of bony tubercles inside the nostrils that guide the shockwaves. Aircraft designers stole the idea. You need to be grateful to peregrine falcons every time you get on a passenger jet.

A further problem for the diving falcon is sight. The bird needs to see if it is to hit the target, and it also needs to look after its eyes in stressful circumstances. The answer is an extra eyelid, which opens and closes horizontally rather than vertically. It's called the nictating membrane and many different creatures have them. It's not necessarily an adaptation for speed, but that's its main function for falcons. A peregrine in a stoop blinks rapidly with its nictating membranes, and this clears the eyes of debris and keeps them moist in the fierce conditions of this headlong dive.

A peregrine is built for speed and power and ferocity. As a result, it's been a symbol of male aggression throughout the ages. So best keep quiet about the females. Because they're the really fearsome ones. They're about a third larger than the males.

You need to be grateful to peregrine falcons every time you get on a passenger jet

Here's how you can invent flight. Cut a vertical strip a couple of inches wide from an A4 sheet of paper and hold it lengthways, just below your lower lip. Now blow. The paper will rise. It will do so because you have created a pressure difference.

Flight happens when air moving across both sides of an aerofoil surface—a wing—creates a pressure difference. An aeroplane does this by driving itself forward. When the air is passing over the wing fast enough, the pressure difference is created and the plane has lift and can take to the air. Higher pressure beneath the wing pushes it up; lower pressure on the upper surface pulls it up; about two-thirds of the lift comes from the pull. Birds create the pressure difference by moving their wings through the air. The downbeat moves the wing forward and downwards, providing most of the thrust; the upbeat, which comes after a loop at the bottom of the swing, moves the wing up and backwards, minimizing drag and still providing a little thrust.

Birds that depend mostly or entirely on powered flight have smaller wings than birds that rely more on gliding and soaring. That's because they are making different aerodynamic demands. So which is the fastest bird in sustained straight-and-level powered flight? It's the start of a good argument, with ducks and geese and wading birds filling most of the top places, with speeds of around 40 mph (64 kph).

The eider, a sea duck that can be found off the coasts of Britain, is probably the champion, with a speed recorded at 47.2 mph (76 kph). It's a chunky heavyweight of a duck, and of all birds it has the greatest weight relative to wing area—the greatest wing-loading, to be more technical. This power enables it to deal with the high winds of the sea-going life.

A wren is so small and so light that flight is almost nothing to it. But as birds get bigger and heavier they begin to explore the limits of their design. The mute swan, the commonest swan found in Britain, is the second heaviest of all flying birds. They're seriously big: it follows that flying is a seriously big deal. They like a runway to get airborne, pattering along the surface of the water until they have enough speed to make flying possible. They prefer to land on water, water-skiing in on huge black feet. They sometimes come down in small, fenced town gardens and can't take off again (the RSPCA is accustomed to doing the rescue job on these occasions). The biggest noise a mute swan makes is the great whoomph of its wings in flight—'the bell-beat of their wings', as W. B. Yeats had it.

The overall winner of the weight-carrying fliers is the kori bustard of Africa, which weighs in at 19 kg (41 lbs). Then comes the mute swan at 18 kg (39 lbs), and in bronze medal position is the Andean condor at 15 kg (33 lbs). These are birds that dwell out on the far reaches of possibility.

Gliding is cheap. You need no power. Gliding is the way that a heavier-than-air machine or a creature equipped with an aerofoil surface can travel forward though the air without thrust. The idea of a glide is to cover a greater distance horizontally than you do vertically. You can see the principle at a rather basic level in the flying squirrel, which uses a membrane of skin to glide from tree to tree.

The all-bird champions of the glide are the albatrosses, all twenty-one species. The wandering albatross is the best of them all, with a wingspan of more than 3.5 metres (nearly 12 feet). These long, thin wings are mimicked by the gliders that humans have created. An albatross's wings allow the bird to cover 600 miles (1,000 km) a day, without so much as a flap.

A glider operates on the same aerodynamic principles that make

left:
THE WANDERING
ALBATROSS,
all-bird champion
of the glide.

powered flight possible: for both, you need a flow of air over the wing. A glider can do this by facing into the wind, which seldom ceases in the Southern Ocean, and by descending under gravity. Gliding efficiency is measured as a ratio of forward travel to distance lost in height. A ski-jumper has a glide ratio of 1.13:1, a metre and a bit of travel for every metre of height lost. A flying squirrel operates at a touch less than 2:1. A wandering albatross has a glide ratio of 22:1.

An albatross's wings lock at the shoulder, which means that it can keep its wings outstretched with no muscular effort. Its heart rate when gliding is not much more than it is at rest. An albatross is acutely aware of airspeed, as every glider must be. The complicated tubes in his beak—unique to the group that contains albatrosses—measure speed very accurately, in the manner of the pitot tubes on aeroplanes.

There is a payback. Albatrosses are not much good at powered flight and calm days leave them grounded—or, rather, sitting on the sea. But windless days in the Southern Ocean are as rare as albatross teeth.

Gliding will only get you so far. 'That's not flying—that's falling with style,' as Woody said to Buzz Lightyear in *Toy Story*, and it's a fair description of gliding. You don't get anywhere as a glider—or in a glider—unless you can soar. That's to say, gain height without recourse to the banal expedient of flapping your wings.

Albatrosses do this by using the wind that gets deflected upward from the waves. They use a combination of slope soaring and dynamic soaring to get a free lift. They dip down to the waves, ride the rising air back up again, continue until they need more height, whereupon they dip and rise again. Their flight across the ocean is a languorous rhythm of ascent and descent.

The champion soarers among the birds are the vultures. There are different groups of vultures in the Old World and the New, and perhaps the Andean condor takes the prize here. They use broad wings with the serrated tips stretched out like fingers. The fingers reduce air turbulence

Albatrosses use a combination of slope soaring and dynamic soaring to get a free lift

and also the stalling speed. The result is that a soaring vulture doesn't need much forward speed to stay aloft.

If you can soar, you can gain height without using much energy. Much fuel, if you prefer, which for a bird means food. Wind deflects upwards from cliff faces and mountain ridges and provides a soaring opportunity. Warm air rises from the ground in columns, called thermals, which is why soaring birds fly in tight spirals, very gradually gaining height. It doesn't happen with any great hurry, but it's highly fuel-efficient. And that explains why the best soaring birds—vultures and eagles—tend to be associated with warm places that have reliable thermals, and/or with mountains that have reliable updraughts. They can't live without lift.

No bird flies like a hummingbird. They change all the rules. When it comes to manoeuvrability they are the clear champions. It's been estimated that they spend 90 per cent of their flying time in the hover: and that's one of the hardest of all manoeuvres. A hovering hummingbird looks as if it were pinned to the air like a butterfly on a card.

They can fly backwards at up to 15 mph (24 kph). Flying sideways and ascending vertically is easy for them. They've been observed performing a somersault between flowers. They really don't fly like birds at all: they fly like insects, and that's not just an illusion that comes from their fascination with flowers.

Hummingbirds have a shoulder pivot that enables them to shift the wing through 180 degrees, and that allows them to move their wings in a figure-of-eight pattern. This way they can get as much value on the up

stroke as they do from the down stroke, like a dual-action pump. In other words, hummingbirds are more like helicopters than aeroplanes. Flight is not the same thing for them as it is for all other birds. The flight muscles of a hummingbird—the pectorals—make up 25 per cent of the bodyweight. They can beat their wings 1,000 times a second.

This is the most demanding lifestyle of any bird. Flight is hugely expensive in terms of energy and fuel requirements, and a hummingbird is perpetually just a few hours from starvation. They must keep topped up on high-energy nectar at all times. They spend only 15 per cent of the day in action: any more would be physically impossible. At night they sink into an energy-saving torpor that's close to hibernation, dropping their body temperature from 41 to 21 degrees Celsius. But when the sun gets up they're hard at it again, helicoptering from flower to flower with a wing noise you can hear at a distance.

Aerobatics is a technical term for silly flying: doing stuff with no point to it, doing stuff just because you can. Some human pilots do all kinds of crazy showy-offy stuff, with stall turns, loops, rolls and spins. And if you thought that the brutal practicalities of life in the wild would put such nonsense out of reach, you'd be wrong.

You often see jackdaws performing a kind of Ferris wheel move in big winds, using very little energy but surfing the power of the wind for no apparent purpose whatsoever. Seagulls often seem to be cruising the cliff tips for the pleasure of their own mastery.

But when any living thing does something completely crazy and with no apparent practical purpose at all, then it helps to have a dirty mind and think about sex. At least first. In courtship displays, some sex-mad birds like to show off their mastery of silly flying.

Meadow pipits parachute; redshanks perform a kind of stutter flight; marsh harriers dance together in the sky and pass food from one to the other as they do so. But I'm inclined to give the prize for aerobatic mastery to the lapwing. They have broad, floppy wings that make them versatile and buoyant. And they defend territories and attract mates by means of aerobatics, all the while accompanying themselves with shrill oboe-squeals. They rise and fall, swoop, dive and plunge. They will loop the loop, and then finish off with a belly-scraping low-level run, like a crop duster. They will make a nosedive so steep you're convinced they're going to land beak-first in the grass, but they pull out at the last microsecond and go into a super-floppy sequence of movements called the butterfly flight. After that they will oscillate from side to side in the wigwag flight. It is at the same time majestic and daft: glorious and futile.

facing page:
A TOPAZ
HUMMINGBIRD.
When it comes
to manoeuvrability,
hummingbirds are the
clear champions.

From afar lapwings are strongly black on top and white underneath, and when they go into one of their aerobatic routines they strobe dismayingly, catching the eye from vast distances, daring peregrines to come and have a go. It's wonderfully crazy and utterly reckless—but it's only by this folly that a male lapwing can fulfil his biological destiny: hold territory, win a mate, keep hold of both, breed and become an ancestor. Birds show us that folly is an essential part of life. Without it we are, quite literally, unable to carry on.

Their numbers are steeply declining, but it's still possible to see 100,000 starlings in the air at once

Why do they do it? The inevitable question at a starling murmuration: those occasions when starlings gather in thousands to perform their dusk dance and win my award for formation flying. Their numbers are steeply declining, but it's still possible to see 100,000 starlings in the air at once.

The question of how they do it is slightly easier. Starlings are small. The message from brain to limbs—wings count as limbs—has a very short distance to travel, so it's easier for a starling to react in a tenth of a second than it is for, say, an Andean condor or a mute swan.

It looks like co-ordinated and choreographed movement—as if each starling were obeying the will of an unseen conductor or air traffic controller—but in fact it's nothing of the kind. They're able to do it because their reaction time is so brief: and in a way, that's a still greater miracle. The idea that they are controlled by a single will is a wonderful illusion for the watching human to savour: but it doesn't reflect the reality for starling-kind.

There are clear advantages in togetherness. A prey animal can always find safety in numbers; it's hard for a peregrine or a sparrowhawk to pick a single bird from a large flock. Even when a predator succeeds in doing so, the odds for each individual in the flock are still pretty good—after all, it's highly unlikely to be you. Any individual is safer when he is one of a large number. There are also advantages in pooling information about food sources; starlings will often forage twenty miles away from the roost. Numbers are also warming on cold nights. For all these reasons it makes sense for starlings to gather in big numbers.

But that doesn't explain why they fly round and round and make a great series of glorious patterns across the sky. Here the objective and cautious speculations of scientists fail us. We are forced to try the imagination and joyous speculation of mere writers.

I see a murmuration as a great celebration of the joys of togetherness. Flying in a murmuration reinforces the bonds between birds: fixes the notion that for as long as winter continues, any one starling is only truly himself when he is part of a crowd. This gathering in numbers is a survival strategy: the murmuration is its expression. Experiencing the joy of being part of a murmuration ties you in more powerfully to the best possible method of getting through the winter, helping you to carry on into the spring to breed and become an ancestor.

A swift spends the first three years of its life in flight. All the time. No perching, no resting on the ground, no holing up at night. Once a swift is fledged and leaves the nest, it won't be in contact with anything but air

above:
THE SWIFT,
all-round flight
champion of the birds.
The air is their home.

until it's time to breed three years down the line. In the course of this time, it will fly to Africa and back every year.

A swift is the all-round flight champion of the birds. The air is its home. They sleep on the wing in huge dreamy circles far out of sight of us earth walkers. They even mate on the wing, screaming hard as they tumble through thousands of feet: a must for the reincarnation wishlist.

For brief periods they are also the world's fastest bird in straight and level flight. They can't beat an eider for sustained speed, but when they're in their screaming parties, racing along streets below the level of the rooftops, they have been clocked at 69.3 mph (111.5 kph).

Their feet are tiny: just big enough to cling to a vertical surface. It's not true that a swift can't get back up into the air once grounded. It's a struggle and rather a comedown, in every sense, after the glorious swooping flight, but adults can manage to hump themselves back into the air. Young birds just fledged, lacking the experience of their elders and betters, have more of a problem, and are sometimes grateful for a helping hand.

So for sheer mastery of the air, I'm giving the title of champion of champions to the swifts: the same species that screams over the towns of Britain and brings the summer back with such style.

A feather has nothing to do with flying. Not at first, anyway. And that's a rather troubling thought. Because a feather is surely the ultimate device for a large-scale flying creature. A feather is immensely strong and immensely light, stiff yet perfectly flexible, and in combination they make an ideal aerofoil surface to create the pressure difference required for flight.

We'll deal with feathers in more detail later (see pages 180–95), but let's pause for a moment to consider that feathers didn't actually evolve for flight. Flight was just a lucky accident. We have 10,000-plus species of bird on this planet, living their own lives and illuminating human lives, and almost all of them do both by means of their ability to fly—and it came about because dinosaurs felt the cold. Flight is a by-product of insulation. Some species of dinosaur evolved feathers for thermo-regulation, an example being *Dilong paradoxus*, related to *Tyrannosaurus rex* but living 60-odd million years earlier. They weren't fliers, they weren't even half-and-halfers attempting to fly. For them, feathers were nothing to do with flight. The feathers developed from the reptilian scale as a way of staving off the cold. These feathered dinosaurs were theropods—and their experiments in insulation filled the air with birds.

How did feathers make the transition from overcoats to wings? It's *Just So Stories* time. Take your pick. Either the theropod dinosaurs found feathers increasingly useful when jumping up to escape from predators (getting higher and travelling further as the arrangement of feathers improved) or when jumping down. Either way, feathers led to true flight, and true flight led to the conquest of the air. Flight changed everything.

Flight is the most extraordinary and, if you like, the most unnatural way of getting about. Every inch you travel is a battle against gravity. And flight starts at the basic design. Just about everything in a bird's anatomy is an adaptation for flight—even if, like feathers, they didn't initially evolve for fight.

Birds have hollow bones. The breastbone is a keel, to which the great muscles for flapping are attached. The system is powered by a mighty four-chambered heart. Birds have a breathing system that is far more efficient than the in-out method we mammals use: one in which hollow bones and air sacs are constantly used to flush out stale air. To fly you must be a spendthrift of energy. Flight is highly expensive in terms of energy, but the bird's breathing system gives it a constant supply of the freshest possible air. A human marathon runner would kill for such an advantage: and having found it, would be unbeatable.

There are other devices to cut down on weight. Birds have no bladders; they let go at once rather than carry the stuff about with them. There are no bones in the tail. A bird has no teeth and no nose; a beak is, like your fingernails, made from keratin, which is strong, pliable and light. The forces of evolution can take it and mould it like putty, so that a toucan can fly with his glorious fruit-picking beak sticking out improbably before him. He looks as if he should nosedive with the first flap, but his beak is so light he can fly directly and powerfully to wherever he needs to go.

Birds have a breathing system that is far more efficient than the in-out method we mammals use

There are about forty species of flightless bird living today. These include the giants like ostriches and emus, and also penguins, which use their wings to fly through water. Kiwis evolved to fill the niche usually filled by night-foraging mammals. They did so in New Zealand, where there are no native terrestrial mammals. There are a few flightless water birds, and some rails—birds like coots and moorhens.

In other words, flightlessness is an exception among birds. For some, like the dodo, what was once a sound evolutionary strategy became a liability when humans arrived on their inaccessible islands and changed their world. Other flightless birds thrive and prosper.

That leaves us with all the other species of bird that can fly: as near to 10,000 as makes no difference. All but the forty-odd can do the thing that humans always long to do, always have longed to do and always will long to do.

Some fly superbly, other can just about get up into the air when there is no other option. But they all have the ability to leave the earth behind them. They all live in three dimensions when we humans must be forever content with two. All flying birds seem that little bit closer to heaven than us. And that is why birds—rather than our fellow mammals—are the most studied non-human animals on earth, and the most celebrated in mythology, painting, poetry and story.

Birds are what we would like to be.

2

Mine eyes have seen the glory

W e're mammals, you and I, so why don't we go out mammal-watching? Because we wouldn't see any, that's why. We hardly ever see non-human species of our fellow mammals in the wild.

True, you can see plenty of mammals on the vast open plains of sub-Saharan Africa, but elsewhere in the world large mammals are less thick on the ground, and even where they're about they're infinitely harder to see than birds. In Britain, it's not easy to get a good view of any wild or non-human mammal, bunnies and grey squirrels apart.

Britain's commonest carnivores are foxes, stoats and weasels. Badgers and otters are about in reasonable numbers. But we hardly ever see them (urban foxes excepted), because they're highly skilled at keeping out of our way. There are otters on the bit of marsh behind our house: I've seen them maybe half a dozen times in three years, mostly at dawn in spring. Our fellow mammals use the cover of night and the very early morning. That's partly to dodge humans and partly because they're comfortable at the times when humans aren't.

But birds are always visible. I may only have seen otters a handful of times on the marsh, but I see birds of prey more or less every day: kestrels, sparrowhawks, buzzards and marsh harriers are not only present but easy to see. Hardly surprising, then, that when we look for the wild world we look for birds. Hardly surprising that across the centuries, birds have meant more to us than any other form of non-human life. Hardly surprising that birds are the most studied creatures on earth.

previous page:
EMERALD BIRD OF PARADISE. Most species of bird of paradise are found on the island of New Guinea, where there is a distinct shortage of predatory mammals.

If you want to be a mammal-watcher, you must resign yourself to being a turd-watcher. You don't often see the animals themselves; the best you can do is to record the signs they leave behind them. I have often found otter turds around the marsh and, for that matter, in the garden. It's always good to find an otter turd, but it doesn't touch my heart.

Otters like to leave turds in prominent places. In fact, turds play a massive part in the way they see the world. When an otter leaves its droppings under a bridge (it's called a spraint) or on some glaringly obvious spot on the riverbank, it's not because it was caught short. It's a matter of policy. A turd is not just a way of disposing of undigested food; it's also a signal, a noticeboard, a potted biography of the otter that laid it. When an otter comes across a spraint he or she will give it a good sniff and by doing so will learn a great deal about the sprainter.

The otter can tell the sex and age of the sprainter and get a good idea of the sprainter's potential as an adversary or mate. A dominant male's spraint is a message that's he's in charge, while an oestrus female will leave a message that can be interpreted as a come-on. Turds are a very effective way of transmitting information for animals that don't bump into each other very often. And I'd be inclined to bet that otters, with noses several hundred times more sensitive than our own, will gather a great deal more information from that single delicious sniff than we scent-impoverished primates can possibly imagine.

But you need a nose.

Oh, we're not too bad at smelling. A smell or a taste—more or less the same thing—can trigger a thousand memories, enough to inspire a novel in seven vast volumes, as Proust demonstrated. But *À la recherche du temps perdu* would have been a lot longer had Proust been a dog, and he would probably have preferred to call it *À la recherche des odeurs perdues*, and to have written the damn thing in smell symbols rather than sound symbols. We humans have five million or so olfactory receptors. A rabbit has about 100 million and a dog 220 million. That's not so much a new sense as a new world: an alien way of understanding life, one that makes possible an entirely new way of living. A dog's fascination with a much-pissed-upon lamp-post becomes understandable. It's like your inbox relayed to you in full colour with a soundtrack by Beethoven.

Most mammals live in a Technicolor world of scent, while we humans have noses that only work in black and white. Furthermore, few of us train our noses to get better; wine-tasters and tea-tasters are exceptions. Most of us are inclined to ignore the majority of information that comes through the nose. Women consistently out-perform men when it comes to the finer discrimination of smells. It's been suggested that as gatherers and preparers of food, especially food for children, they needed a more acute sense of smell if their children were to survive and they were to become ancestors. But we're all amateurs when it comes to most of our fellow mammals. Bears are reckoned to have a sense of smell seven times more efficient than dogs, enabling them to find food hidden underground. We humans aren't on the same page.

We can't get around in the dark with the aid of our noses. We can't understand the doings of our own kind by sniffing. We are—before anything else—visual animals. That's why we love the day and fear the night. That's why daylight white is a good colour in our eyes and night-time black is sinister. Daylight brings us safety and security because during

Most mammals live in a Technicolor world of scent, while we humans have noses that only work in black and white

the day we can see. During the day we know where we are. During the day we can understand the world.

And so can most birds.

Birds don't really bother with smell. Albatrosses can smell, and that helps them to find food across the oceans—wandering albatrosses have been recorded latching onto a food source 12 miles (20 km) distant. New World vultures, like the Andean condor, can also smell, and that guides them to food. Kiwis, uniquely among birds, have nostrils at the tips of their beaks rather than at the root, and they use this arrangement to sniff out worms, which they find by poking their beaks into the earth.

But most birds understand the world by their sense of sight, and there they are at one with us humans. Seeing is believing, after all: so most birds start their day in the light and end it when darkness comes.

They can also, as said before, fly. That means they don't have to hide from us humans. The sparrow that forages for crumbs at a picnic knows that he can take to the air much faster than you can pounce, so he has no need to hide. The gulls flying past, the blackbird in the bush, the crow at the top of the tree: they have no objection to being seen. They know they have the wings of us.

If you want to see a hunting mammal—say, a badger—you must know the right place and get there before nightfall and sit there, very quiet and still, with the wind blowing from the badgers to you, so they will be unable to smell your presence. And if you do so for a couple of hours, you're quite likely to get lucky; the badgers will emerge from the sett and start snuffling about. But if you want to see a hunting bird, all you have to do is drive along a motorway.

No wonder birds are the most studied and the most watched group of animals on the planet. Even a human can't miss them.

Most of our fellow mammals can out-smell us. But we have the advantage when it comes to sight. Most mammals have two kinds of light-receptive cones in their eyes; all of us Old World primates have three. Lemurs, monkeys, apes and humans see a world that most mammals can't begin to imagine. We see in colours beyond the experience of a dog, a horse or a cow: a world beyond the possibilities of their minds.

It's important to realize here that colour is not a fixed and measurable thing. It's not an inherent property of the thing you are looking at. Colour is the response of the nervous system of the animal that's doing the looking: the animal whose eyes are receiving the reflected light and whose brain is processing the information. A colour-blind human being is not seeing things wrong, he's seeing things differently.

We humans (mostly) see the world as a three-coloured place, and when we make colour televisions we transmit this artificial world in red, green and blue lights. Everything we humans can experience in terms of colour vision can be re-created by mixing these three colours. That's good enough for us: but it's not the end of the story. If you take LSD you are likely to experience colour in a different way, and see tangerine trees and marmalade skies. Such an experience is no less valid than the normal perception of green trees and blue skies. Colour is not an objective fact.

What birds can see is beyond our experience, beyond our imagination

Birds are visual animals: they can see better than, or at least quite differently from, all of us mammals. That's because birds have *at least* four different types of light-receptive cones in their eyes. They see colour in four dimensions. They can see ultraviolet light. We call it ultraviolet because violet is the limit of our vision; shorter wavelengths are beyond us, but not beyond birds.

It's clear, then, that birds can see colours beyond the limits even of us trichromatic humans. We used to think, in our vainglorious way, that we humans had the best colour vision on the planet and that birds were lucky enough to share it. Now we know that they are way beyond us, and that the colourful birds that bring us such joy are, in a bird's eyes, more colourful than we could ever dream or hope of experiencing. Male and female blue tits look the same to us, but they're glaringly different to a blue tit. That's because of the ultraviolet light reflected from the crest, which can be experienced by a tetrachromatic bird but not by a trichromatic human.

What birds can see is beyond our experience, beyond our imagination, beyond the possibilities of our imagination. But even in the muted form that birds' colours come to us we find something glorious, something that daily lifts our hearts. We are sight animals before we are anything else, and so we respond to the sight of a bird far more powerfully than we do to the scent of a fellow mammal.

Birds may be a little beyond our scope in terms of colour, just as they are beyond our scope in their ability to fly. But that doesn't mean we are unable to respond to them as fully as humans can. And in the form of colour that flies, birds bring us the most vivid understanding of the non-human world that we are capable of experiencing.

You may have been lucky enough to see this bird in Britain. It's a spectacular example of colouration. The sides of the neck are glossy purple and green, set off with two prominent white patches. The pale yellow eyes add drama to the head. The front is a rich pinky-purple, merging subtly into rich textures of mushroom and mauve. Each wing is emphasized by a white crescent. Naturally, the bird loves to show off such luxuriance of colour and design, and it does so in an extravagant display flight. First it will take off and attract attention with a double clap of its wings. Then it will perform a stiff-winged diving glide designed to stress the perfection of its colour scheme: a languorous passage across the sky that makes it plain that this is one hell of a good bird, and you'd be well advised either to keep away or marry it on the spot, depending on your gender.

This gorgeous bird is the wood pigeon. Jeremy Mynott, in his excellent book *Birdscapes*, unilaterally declared the wood pigeon to be Britain's least charismatic bird. You never hear a birder boast of 'having' a wood pigeon. Even when seeing them in prodigious numbers in their foraging flocks outside the breeding season, few people break the rhythm of a walk to look at wood pigeons. They are regarded as the dullest of the dull.

And that asks two questions at once. First, why don't we look at the wild world with a little bit more care and attention? If we did, we would see even more wonders than we do already. And secondly, if this sumptuously coloured animal is regarded as the dullest bird in Britain, how gorgeous must the rest of them be?

Some British species would be regarded as the most wonderful birds in the world if only they had the decency to be rare. Some of our commonest birds would seem to us staggeringly lovely—as lovely as a red-flanked bluetail or a Siberian rubythroat—if we only saw them once or twice in a lifetime.

Take the cock chaffinch. Catch him in a good light and he's staggeringly handsome. It's the audacious combination of the powder-blue head and nape with the rich, deep pink of the face and tum. This bird is outrageous, but because chaffinches are easy to see all over Europe, being great lovers of parks and gardens with a marked tolerance of humans, they have no mystery. We take their fine colours for granted: just part of the background.

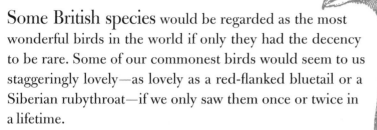

The male mallard is probably the most spectacular of all the birds we take for granted. He's a riot of colour, wildly over the top. His front end is a daring combination of bottle-green, deep purple-chestnut and Van Gogh yellow. The green is for the head—which in some lights shows as a deep bluey-purple—the chestnut for the chest and the yellow for the beak, all set off with a narrow but unmissable white collar. The wings are stressed with a tiny patch of shining blue, called the speculum, which is also carried by the drabber female. When a mallard drake opens his wings the patch of blue becomes long and wide, emphasized by a white stripe. As he flies away from you he will flare his back end and show you a dark tail with crisp white outer feathers.

Wood pigeon, chaffinch, mallard: three of the most frequently seen birds in Britain and Europe: birds we never worry about, birds we never enthuse about, birds we seldom even mention. I know no great poems to the beauty of pigeons. The great nature poets seem to have left these birds uncelebrated. Their loveliness is there for us all to revel in, if only we could be bothered. But mostly we can't, and not least because there are so many other birds that effortlessly command our attention and fill our hearts and minds.

I know no great poems to the beauty of pigeons. The great nature poets seem to have left these birds uncelebrated

The kingfisher is the bird of dreams: the bird most people long to see. They want to see a kingfisher for the impossible luxuriance of its colours: electric-blue on top and rich salmon-pink underneath. The colours seem almost gratuitous, as if there were no reason for such colours save love of life.

I bet that more than half the people who say they have never seen a kingfisher have actually seen one—but thought it was something completely different. What, you ask? Seen a kingfisher and mistaken it for something else? Surely this is the one bird no one could mistake: a flying furnace of reckless colours that burn themselves into your retina. If you see a kingfisher, you must know for certain that you have done so.

Not necessarily. What's that tiny black bird fizzing so low across the water, in sight for no more than a second? Probably a wren, or one of those little brown jobs at the back of the field guide—drab birds that it's easier not to bother with. But that was your kingfisher. You hear a kind of squeak, a bit like a doggy toy. Another black flash. And gone.

Kingfisher.

Birds have two ways of expressing colour. One of these is by means of pigment—like paint—and that's the way we humans tend to understand the idea of colour. But the other is pure light. Many bird colours are the result of the refraction of light, and that's another of the extraordinary things that feathers can do. Feathers can refract light like a prism.

You pass a puddle in a city. A pretty ordinary experience. And then the angle of your gaze shifts, and the puddle is lit up with impossible rainbows of colour; in that puddle you can see kingfisher blues and kingfisher pinks. There's a drop of oil in the water; it floats on the surface and creates an impromptu refractive device. In a dirty puddle you find an approximate image of heaven.

Thus drab birds can be suddenly aflame with iridescence when seen in a good light. A springtime starling is perhaps the most startling example. Here is a bird with a reputation for drabness, and yet in certain lights it explodes into a heavenly creature of glowing purple, lit with deep blues and greens, all of it spangled with gold. Perhaps its fellow birds always see a sumptuously coloured bird, but to human eyes this is one of those movie make-over moments: the hero removes the heroine's glasses and tells her she's beautiful. To which the only correct reply is 'I know'.

Some African starlings are gorgeous at first look: the species called the splendid starling and the superb starling are not overstating their case. But any starling, seen with the right eyes or in the right light, is as splendid as it is superb.

There is an effect called 'scattering', which also involves refraction, operating on tiny air pockets within the feathers. These produce the colours in one of the flashiest groups on earth: the kingfishers, bee-eaters and rollers.

'The sight of a feather in a peacock's tail, whenever I gaze at it, makes me feel sick'

Birds use three kinds of pigments to express colour. The first is melanin, the same pigment that makes humans go darker in the sun. Unsurprisingly, it colours feathers black and brown, and is particularly helpful to a bird because it adds strength to a feather.

Many fishing birds are predominantly or entirely white. That way, they are close to invisible to a fish looking upwards towards the light. But a number of white fishing birds have black wing tips. The tips are where you need most strength in a feather, particularly in gliding and soaring flight. And the air is full of examples: gulls, pelicans, gannets and albatrosses. An adult wandering albatross is pure white—except for those crucially strong black wing tips.

Carotenoids come from plants, and birds can acquire the pigment by eating either the plant or animals that have eaten the plant. The yellow on a goldfinch's wing comes from carotenoids; reds and oranges can come from the same source. Flamingos are pink because of the carotenoids they ingest. They turn white when deprived of them, as you sometimes see in zoo collections.

The third pigment comes from porphyrins, which are modified amino acids, and they can provide reds, browns, pinks and greens. So many different ways of making colours, and so many birds make such extravagant use of them.

Why?

Peacocks made Charles Darwin want to vomit. 'The sight of a feather in a peacock's tail, whenever I gaze at it, makes me feel sick,' he wrote uncompromisingly.

And no wonder.

The function of a peacock's tail is to be beautiful. There is no other explanation. Peacocks carry this vast train of wondrous and impossible colours for no reason other than its beauty. They take the notion of iridescence to its logical extreme: refracting the light in blues and greens that dazzle the eyes of us poor humans. How much more wonderful do they look to a peahen, I wonder?

The key to Darwin's thesis in *The Origin of Species* is that every modification passed on from one successful animal to another gives that animal a better chance of surviving and becoming an ancestor. That tail seems to contradict every argument that Darwin put forward.

The tail doesn't help a peacock survive. One the contrary, it's a rather noticeable handicap. A peacock with a train is far more vulnerable than one without: far more likely to get grabbed by a predator. So why does beauty thrive in the face of the practicalities of survival?

Darwin had a mind like a rock crusher. Give him a problem and he would never give up on it, slowly grinding it to powder under the inexorable force of his thoughts. So it was with the vomit-making peacock.

Sexual selection. An individual isn't just competing against other species for resources and for survival; he or she is also competing against his or her own kind. And within a species, an animal is not just competing for food and space; he or she is also competing for access to mates. You can't become an ancestor without a partner or partners. So how do you choose?

In some species of birds, the male attracts the female by being much brighter in colour. Many put on a performance to show off these colours, these important points of difference. A male bird would be much safer in drab colours, skulking about in cover, but he wouldn't get any females. Hence the need for colours.

You can find this in many different species: most obviously—because of their high visibility—in many species of duck. The gorgeous male mallard has already been mentioned in this chapter: the female is dull brown with that discreet flash of blue—the speculum—in the wing.

And peacocks have this to an extreme degree.

'The sight of a feather in a peacock's tail, whenever I gaze at it, makes me feel sick'

Sexual selection is now accepted more or less universally. It's part of the way that life works, one of the reasons why life is the way it is. But it was a while before it became scientific orthodoxy, because there is one small but highly significant point hidden behind the dazzling nature of a peacock's tail.

Female choice.

The competition among males is not set down under male rules. It is a competition judged entirely by females. In other words, female choice has dictated the nature of many of the species that live on earth. Some might feel uncomfortable with such a fact today—it was a notion almost horrifying to the Victorians.

But it's the right answer.

The birds of paradise took the idea of sexual selection and made it their own. They are related most closely to crows: and yet they have evolved into forty-one of the most extreme species on earth: dazzling colours, extravagant plumes and outrageous displays involving dances, extraordinary noises and even perching upside down.

Male birds of paradise are shatteringly different from females, so much so that they look like members of a different genus or even family. The females are dull brown things for the most part, the better able to hide while incubating eggs. Most species of birds of paradise are found on the island of New Guinea, where there is abundant food and a complete lack of predatory mammals, apart from those introduced by humans.

This has given the birds leisure time, as it were: the opportunity to compete for mates on grounds other than their skills in foraging and staying alive. After all, anyone can do that. So the birds have flourished and developed into these exaggerated forms, all driven by the tyranny of female choice. These drab-looking things, looking on with an air of tolerant amusement as the mad males, desperate with lust and packed to the gunwales with testosterone, perform for their pleasure, are the bosses of the show: the driving force behind this bravura passage of evolution.

The extravagant names humans have given to birds of paradise reflect their extraordinary nature: King of Saxony bird of paradise, Princess Stephanie's astrapia, the king, the emperor and the magnificent bird of paradise. The birds of paradise were one of the revelations of Alfred Russel Wallace, the great explorer and, with Darwin, co-discoverer of the theory of evolution by means of natural selection. These birds have also been a lifelong fascination for Sir David Attenborough, whose *Attenborough in Paradise* programme brought us footage of the birds in action in a manner that dazzled the eye and frazzled the brain. In the final piece to camera in this programme, Attenborough says:

> Wallace's emotions on discovering such marvels must surely be echoed by all of us who follow him. This is what he wrote:

> 'I thought of the long ages of the past during which the successive generations of these things of beauty had run their course. Year by year being born and living and dying amid these dark gloomy woods, with no intelligent eye to gaze upon their loveliness, to all appearances such a wanton waste of beauty. It seems sad that on the one hand such exqui-site creatures should live out their lives and exhibit their charms only

in these wild, inhospitable regions. This consideration must surely tell us that all living things were not made for man. Many of them have no relation to him. Their happiness and enjoyments, their loves and hates, their struggles for existence, their vigorous life and early death, would seem to be immediately related to their own well-being and perpetuation alone.'

Attenborough lifted his head to the camera with that easy intimacy that has marked him for more than sixty years of broadcasting and concluded: 'Indeed so.'

But what about kingfishers? Males and females are as beautiful as each other. The males do not compete for gaudiness: females are just as lovely. They clearly aren't choosing males on the grounds of their extravagance.

The answer seems to be that it's a good way of recognizing each other. It's important for humans not to copulate with chimpanzees. They are our nearest relations—we have getting on for 99 per cent of our genetic material in common—but mating with each other across the species barrier would be futile. And we don't. It's a problem that's easy to avoid. We are both strongly visual animals, like all other Old World primates, and so we can recognize our own kind with a single look.

If you want to become an ancestor, you're wasting your time unless you mate with an animal of the same species. So all species send out signals to each other that indicate their relatedness. A dog, no matter how humanized, knows he is a dog and can recognize fellow dogs, no matter how much they may differ superficially. A poodle, a mastiff and a dachshund are all members of the same species, and they know it: presumably not from a quick glance but from a quick sniff.

So kingfishers can recognize each other from the identical beauty of their plumage. They nest deep in burrows so there is no need for the female to be drab and inconspicuous. A kingfisher, when still—and their basic hunting strategy is stillness—is remarkably difficult to see. And as said before, they are only gaudy in certain lights.

Kingfishers are lovely. They're not always visible but their highly colourful plumage is a matter of instant recognition, for both potential mates and potential rivals. This is useful because the birds tend to be strung out

below:
KINGFISHERS
seem to be more
beautiful than they
need to be.

along the length of a river; they don't bump into each other all the time and every day. Colour is a very helpful rapid recognition clue that works well for their way of life.

They still seem to be more beautiful than they need to be.

I've only seen hyacinth macaws in captivity, so I'm not sure if I can really count them: but they must be in the mix for anyone looking for the all-bird champion of colour. There's a fine colony in London Zoo, managed for conservation purposes.

They are more than 1 metre (3 ft) from head to tail tip—the world's longest parrot and the heaviest flying parrot. They use tools (they will use a stick of chewed leaf to keep a hard nut in place while they open it) and that adds another touch of glamour, but their colour is outrageous—an impossibly rich yet subtle range of blues set off with yellow.

They excavate rows of burrows in the sandy cliffs: from a distance the cliff face looks like a man o' war about to fire a broadside

But for sheer bravura drama of colour, you should visit one of the great carmine bee-eater colonies. The most accessible can be found in the Luangwa Valley in Zambia. These birds are hole-nesters, like king-fishers, but they like to do so in large numbers, and they excavate rows of parallel burrows in the tall, sandy cliffs of the river, so that from a distance the cliff face looks like a man o' war about to fire a broadside.

The air fills with the sharp Pekinese yap of these birds as they call to each other. At the start of the nesting season you will see constant little puffs of sand emerge from the holes as the birds excavate and enlarge them, so that it looks as if cannons within have been silently fired at the enemy.

One single carmine bee-eater is way over the top: a deep, rich cherry red, with an electric-blue face set off with a load of black eyeliner. As if that wasn't enough, the blue is repeated on what we naturalists call the bum. They show all this off with a swoopy, glidey flight as they hawk for insects on the wing, and they are finely made things that carry a single long streamer in the tail. Now multiply this by almost a thousand, for when they all take to the air together they turn it into a hallucinogenic overload of colour, challenging your ability to accept reality for what it is.

The science fiction writer Carl Sagan was—at least according to legend—discussing sending material into space in an effort to contact extra-terrestrial civilizations. It was suggested that they send examples of music by Bach. Sagan was against this, on the grounds that it would be showing off. I'm inclined to think that the carmine bee-eater is the class of Aves doing the same thing.

3

Song
of myself

We humans can hear pretty well. We're also quite good at making complex sounds. So we sing. Almost alone among our fellow mammals, we sing long and complicated songs.

On land we have no serious rivals as singers. Lions communicate with roars that carry across the African night with a sound like a giant's belch; hyenas locate each other with a spooky whoop; wolves howl to keep in contact at distance; families of gibbons sing in chorus at dawn as a daily way of reclaiming their own area of rainforest; the carnivorous grasshopper mouse will send out a sustained trill. The tree hyrax—a rodent-like creature that's actually related to elephants—makes an astonishing sound in the forests at night, based around a crazy and repeated screaming. And though all these sounds are wild and inspiring and wonderful, they're not the sort of thing we readily call music.

Whalesong is one of the great discoveries of recent decades. It was first discovered by American servicemen listening out for Soviet submarines in 1952. Since then whalesong has been much studied, especially the songs of humpback whales. Astonishing things have been revealed: a new song, said the cetologist Phillip Clapham, was found to 'spread like a wave across the Pacific Ocean' as different individuals picked it up, learned it and performed it: a clear example of musical culture passed on from one whale to another.

But we've known about this for less than a century. The songs of birds were part of our lives before our ancestors were fully human.

previous page:
NIGHTINGALE.
For caption,
see page 72.

We learned melody from the birds: and thus music became a part of all human cultures. To create music, as we humans understand the term, we set melody against rhythm. And all of us mammals are natural rhythmicians. We humans spend the first nine months of our existence listening to the great drum solo of our mother's heartbeat. We are powered by that immense even-time rhythm; four-four is the most basic time signature that musicians work in, and it's the one that cuts deepest.

But it's my belief that we filched melody from the birds. Listen to a blackbird (you can do so easily enough by googling 'blackbird RSPB'). It sounds astonishingly like a human whistle: relaxed and easy, as if the whistler was leaning against a wall with his hands in his pockets on a warm day. And we humans can imitate this. To do so is a natural impulse.

And it's a tune. It's melody. And it's more or less music as we know it. Across the world you can hear birds every bit as tuneful. The first humans who walked upright on the savannahs of Africa would have listened to the songs of robins and larks, found comfort in them and imitated them. And added rhythm.

The first musical instruments created by humans were bird flutes. These were made of bone, and they were capable of making two or three notes that sounded like the notes of birds. The oldest of all is the Divje Babe flute found in Slovenia. It's 67,000 years old and very fragile, made from the femur of a young cave bear, an animal that became extinct 25,000 years ago.

Other flutes have been found from more recent times, and they are often made from birds' bones. Bird bones are hollow, as we have seen: they are practically a flute already. A five-hole flute was found in Ulm in Germany, and it's about 35,000 years old. The earliest instruments that are not flutes are very recent indeed. They date from 2600 BC: a collection of harps and lyres found in Sumeria.

Surely it would have been easier to make a plucked instrument from a length of gut. Or a melodic percussion instrument, like a xylophone, from chunks of wood. But that didn't fit our first ideas of music. We wanted to sound like birds.

Birds and mammals have different equipment for making sounds. Birds have a syrinx instead of a larynx: a forked organ that some birds use to sing two notes at the same time. The syrinx is named for a chaste nymph who was lustfully pursued by Pan. She hid in the water and prayed that she might be transformed into a reed and so escape Pan's attentions. Her prayer was granted and Pan turned the reed into... pan pipes.

As we have seen, birds also breathe differently from us mammals, with adaptations that help them to fly, for powered flight is highly expensive in energy and requires a constant access to unbreathed, oxygen-rich air. As a bonus, this also enables birds to produce long, loud and almost continuous sounds, and to put on a performance that's way beyond human singers, forever limited by our in-out method of breathing. Virtuoso performers on wind instruments can master circular breathing, a way of seeming to exhale continuously. The trick is done by forcing stored air

left:
A SKYLARK.
The miracle is not that
skylarks can sing forever
on a single breath, it's
the fact that they are
equipped in a manner
that makes continuous
singing possible.

out by pressure of the cheek muscles while inhaling through the nose. Birds can do the same thing without training: it's in the structure of their breathing apparatus.

So when you listen to a skylark singing at the top of its voice while flying steadily up towards the clouds, it's not quite the same as asking you to run a mile while singing the 'Hallelujah' chorus. The miracle is not that skylarks can sing forever on a single breath, it's the fact that they are equipped in a manner that makes continuous singing possible.

Every song says something about the singer. A good burst of song tells listeners that this is a bird in very good shape. It has to be: the wren whose song fills your garden is tiny; you could hold a dozen in your cupped hands. Their astonishing Brian Blessed level of vocal power is a demonstration of the strength of the species—and of the individual.

A bird puts everything into a song. It's a massive explosion of energy, even when it sounds as laid-back as a blackbird. So if you listen to a bird with a critical ear, you can make accurate judgement of just how fit and strong that individual happens to be. And that's what female songbirds do. They judge a male's song just as a peahen will judge a peacock's tail.

Birds hear better than us. Their vision, as we have seen, is more receptive to colour than our own. Their hearing is better at distinguishing very brief passages of sound. Let's go back to the wren in your garden. If you listen to him, you will find that he finishes his song with a powerful sustained trill. It's pretty impressive for such a tiny bird, but to a human ear it's not very interesting musically.

So record it and then play it back slowed down. You will find something completely different. (There are examples on the internet.) I have listened to a recording of a wren's song that lasted 8.25 seconds, and listened to the same thing slowed down to last 66 seconds. And the transformation is startling: from the chrysalis of a dry and rather mechanical trill you find emerging, like a butterfly, a gloriously sweet melodic line. Analysis revealed that the wren sang 103 separate notes in the course of that eight-second song. He was singing at a rate of 740 notes a minute. Guitarists reckon that you move into another dimension of shredding when you reach a speed of 170 notes a minutes. It takes a virtuoso on the instrument to rival a wren.

Does a bird hear each individual note? The easy answer is to say that they must do, or what would be the point of singing them? But we can do better than that. The whip-poor-will is an American bird famous for its three-note call; its name is an onomatopoeia. But if you slow the call down, you will find that there are actually five notes in there.

The mockingbird is a famous mimic, and it likes to imitate many different kinds of birds, among them the whip-poor-will. Slow down a mockingbird's imitation of a whip-poor-will and you find... five notes. In other words, even if birdsong does move humans very powerfully, it's still a bit beyond our scope to hear every note.

Birds mostly make noises so that other birds can hear them. The sounds they make have a meaning and are uttered with a purpose. It's communication. You can call it language if you like, though most philosophers' definitions of language are skewed in a desperate attempt to demonstrate that language is uniquely human: we want to keep the club exclusive. Plenty of evidence has been discovered to make the barrier between human and non-human communication fuzzy. Famous examples include Washoe, the chimpanzee who learned American Sign Language, and the honeybees who inform each other about food sources by means of the famous waggle dance.

Birds are able to communicate with each other through sound, so let's leave it at that. You can break this down into two different kinds of sound, with radically different functions: call and song. Call tends to be the very direct communication of a simple idea, and so the sound is usually brief and straightforward: often monosyllabic, and seldom what you'd call tuneful. The most common calls are those that give a warning of danger—an alarm call—and those that give 'I am here' information, a contact call.

Song is much more elaborate and complex. Song is a way of passing on information not about the world but about the individual singer: how effective he is and how experienced (singing birds are usually males). Not all birds sing. In fact, most don't. All the same, birdsong is a global phenomenon and those birds that do sing have been phenomenally inventive and successful. They have also given delight across the centuries to humans.

Step out into your garden and at once you hear that rattling retreat of a blackbird

Are blackbirds altruistic? Step out into your garden, or take a less well-trodden path in the park, and at once you hear that rattling retreat of a blackbird. So you need to ask yourself: is this altruism in action? Is the bird putting himself at a disadvantage by making this call, attracting attention to himself and thereby handing the advantage to other birds? Is it a reciprocal favour from which everyone benefits, like letting people get out of the lift before you get on?

It's an argument that can chase its tail forever. You can talk about cost-benefit trade-off. You can suggest that by uttering an alarm call the bird is warning his own relatives of danger, thus he is looking after the future of his own genes. You can argue that an alarm call is also a signal to the predator: it tells him that he has been spotted, so there's no point in continuing hunting operations around here. It's also possible that the blackbird's sudden shout is a startle device, one that stops the predator in his tracks for a fragment of a second, enough to allow the blackbird to retreat.

Pheasants are great exponents of the startle technique. They will lie doggo in cover until you are almost on top of them, and then they will break cover with the most fearful din they are capable of. It's an alarm call, and it frequently makes a human jump. By the time you have recovered, the bird has gone clear.

In woodland you will often hear a thin, high-pitched whistle. This is a universal alarm call, one that many species will use to pass on information of potential danger. It's a very hard call to locate, so the bird isn't really giving himself away, and it works from one species to the next, so it's a pooling of resources. It's a kind of Esperanto: a universal language of the hunted.

I'm here! Sometimes with an implied follow-up: where are you? Or follow me! Or, with a chick, feed me!

The calls will be brief, mostly monosyllabic and quite often repeated. They are useful if you want to be found in, say, a woodland canopy, where you can't see for more than a few inches. Birds will call and call back: it's useful information and a form of reassurance.

Of course, this could also be a come-and-get-me signal to an alert predator, so when there's a predator around it pays to shut up. And that means a sudden silence is as good as a shout. Stopping your contact calls is a silent alarm signal, an important warning for all that can hear: especially mates or young.

If you get close to a feeding flock of flamingos, you will find a comic element to their elegance and beauty. From a distance, you are overwhelmed by the mass of pink feathers—all courtesy of the carotenoids, of course—and the long, lovely lines of their bodies. But when you're up close, your ears are filled with a daft quacking: thousands and thousands of birds, all quacking every now and then, so that it sounds like a great rolling thunder of quacks. Bring in a hyena or an African fish eagle and the quacking stops, as if a switch has been thrown: and then the birds take to the air.

Watch a skein of geese flying over: always in echelon, like cyclists in the Tour de France. They do so for the same reason: it's much more energy-efficient to travel in someone's slipstream than to take all the pressure of the air on yourself.

Geese constantly honk as they fly. It's a valuable way of keeping in touch. It is important to know precisely—to the inch—where the next bird is, because that way you keep the formation tight and all except the lead bird can save energy. This is especially useful on migration flights, when geese will often travel at night, but it's useful at any time, because you can't see the bird behind you. I also suspect—though this is pure speculation, so miss this bit if you have a taste for scientific rigour—that the constant honking is a form of group solidarity, all us geese together, and as such it's a powerful encouragement for the geese to spell each other and take turns at the front. Give all the work to a single goose and it'll slow us all down and make us all less efficient. Taking turns is good for everyone. And so they honk.

Contact calls are seldom pretty or elegant or heart-stopping. There's little music to be found here. They are basic survival tools, and if you listen out you will soon learn what species makes what call, and so you will be let into the secret. A jackdaw calls, *Jack! Jack!* And a flock of jackdaws will often all call together, so that I am reminded of a snooker hall. A single game of snooker is a mixture of silences and sharp clacks as ball hits ball. But when there are thirty tables at a snooker club in operation

below:
JACKDAWS.
When a flock of jackdaws flies over, they will jack cheerfully to one another.

at once, the clacking is continuous, so that the entire quality of the sonic experience is changed and it becomes an unending sound that means snooker.

And when a flock of jackdaws flies over, often revelling in their ability to ride the wind, they will jack cheerfully one to another, so that the air is filled with a great rolling and jacking. It's the sound of togetherness and of the birds' own comfort—and, I dare say, joy—in sharing the air with their own kind. At cricket matches the Barmy Army sing their own name endlessly, or at least they do in the evening session when many of them are capable of little else. Jackdaws are doing the same thing.

Call is about staying alive. Song is about making life worth living. Broadly speaking, anyway. If you'd prefer that a bit less anthropomorphic, call is about not dying while song is about living forever. Song is about making more birds. It's about sex and territory, and for many birds they're pretty close to being one and the same thing. Every song every bird ever sang is the song of life.

When a cock bird sings out he sings the song of himself. Sure, the song will tell other birds and human listeners what species he is, but that's just the start of it. He will also tell the world what kind of individual he is: whether he is a bird that a male rival will fear, whether he is a cock-bird that a hen-bird wants to mate with.

A song is an invitation to share genes; to have a punt at the task of becoming an ancestor; to try to make sure that, though your body will die, your genes will live forever.

Song is first a signal: both a come-on and a piss-off. It's a come-on to likely females and a piss-off to rival males. When a great tit sings a ringing, simple phrase in the early spring—often transcribed as *teacher-teacher-teacher*—he is saying that the great-tit niche in this place is more than adequately filled: so if you're a male great tit, move on, and if you're a female great tit, move in.

But the song doesn't just give information about the species and the sex of the singer. It also gives the listener important information about the individual singer. It's hard to sing; it's still harder to sing well; it's very hard indeed to sing loud and sing long. A really good song tells the males that this is not the bird to take on in a singing competition, still less to pick a fight with. It also tells the hens that this is a bird capable of defending a rich territory packed with food resources, the perfect place to bring up a brood of little great tits. In short, the singer is a damn good prospect.

Territory is not the same as property. No doubt human ideas of ownership can be traced back to our atavistic territorial urges, but the great tit doesn't own a stretch of garden in the same way that a person owns a house or a field.

For a start, it's not a permanent thing. Most birds will abandon not only territory, but also any notion of trying to hold on to it for most of the year. The urge for ownership comes with the seasonal urges for procreation, and it's abandoned as the seasons change.

Secondly, the bird can't develop it. He has no rights over it and many other species of bird will hold territories that coincide and overlap with what he considers his own territory. Mostly, that doesn't matter. The wren's song is different from the great tit's, and since he occupies a different ecological niche—different food, different ways of looking for it, different places to look for it—he won't be considered a rival. So the two coexist: you can hear your wren singing at 740 notes a minute in the same corner of the garden as the great tit's *teacher-teachering*. The wren will be in the undergrowth at the bottom of the tree, the great tit will be at the top.

The boundaries of a territory may not be tightly defined, in the legal, fence-building sense in which we humans understand it. Territories can overlap. A great tit may tolerate another great tit on the fringe of his territory, especially when he's busy at the other end. But best not try singing your head off just by the nest hole: that's not so much asking as begging for trouble.

Territory is a looser concept than we imagine if we think of our own homes. But all the same, it's life and death to a territorial songbird. And it is established and held by means of song.

Broadly speaking, there are two kinds of birdsong and two kinds of singers. There are stereotypical singers and there are repertoire singers. These categories aren't exactly hard and fast, but it's a useful idea, especially when you, dear reader, are trying to learn a few songs and tune into the wild world with your ears.

Stereotypical singers sing the same song again and again and again. They never get bored with it. Take a chaffinch, the finch whose lovely

colouring we discussed in the last chapter. Chaffinches have a range of calls, and one of their contact calls sounds pretty much like *finch*. And that gave the name to the whole family of chunky-billed seed-eaters.

Chaffinches are one of the first birds to start singing in the British spring, and they're always welcomed for that reason. It may not be the finest or the most inventive song in the wild world, but it's the chaffinches' own and they give it everything. It's an accelerating succession of notes that ends with a flourish. It's said to be a picture in sound of a fast bowler in cricket: a rapidly quickening run-up followed by the explosive delivery stride. It builds up and then it finishes with a bit of modest style.

And—er—that's it. Once you've learned it you'll hear it right through the spring. It's not overly fancy, but it works for the chaffinch. And to a human ear every chaffinch sounds pretty much the same pretty much all of the time.

I was once involved with some people who wanted to invent a birding app like Shazam: you could make your phone listen to a bird singing and instantly know the name of the bird. I was writing a book on birdsong at the time and it seemed that this was a Great Business Opportunity.

I was a bit of a wet blanket, I'm afraid. Birds are individuals, I said, and they sing as individuals. Even chaffinches. It's not like a record: every one is different. They're not clones. They don't sound like clones. Or machines. Or records.

It never got off the ground, though there are now apps that claim to pull it off: one called Warblr, another called ChirpOMatic. Reviews say they are some way short of infallible. We try to avoid stereotyping humans on the lines of gender or race or profession; we should also avoid stereotyping birds. We should even avoid stereotyping stereotypical singers.

Italian chaffinches sing with Italian accents. There was a time when I made a visit to Rome every other year, always in the spring. On Sunday morning I would take a stroll in the Villa Borghese gardens, and there I would hear recognizable chaffinches that sang slightly but noticeably differently from the chaffinches in my garden at home.

There are regional variations in birdsong. It's a relatively subtle business, but it can't be all that subtle if I can pick it up. It's a bit like hearing your favourite band live: your favourite song doesn't sound quite like it did on the album. You know it so well you can tell the difference: and that makes the experience all the more meaningful. In a way it humanizes the music and makes you more closely involved with the band and their songs. And in the same way, the variations in birdsong bring us closer to the avian singers. In both cases, for the listener, there is a sense of belonging that wasn't there before.

And this demonstrates again that not even stereotypical singers are slaves to their own song. The force, the volume, the exuberance and the number of times they can sing it give important information about the power and effectiveness of the individual bird: and even in a bird that thrives on repetition there is musicality, for that comes in the natural and subtle and inevitable variations.

Repertoire singers raise the game. They are being judged not just on volume and power and endurance: they are also being judged on their musicality. The more complex the song, the more desirable the singer is in the ears of the females. The females are more inclined towards males with greater range and variation: yes, we're back in the realm of female choice. Birdsong delights human ears. Birds sing in order to give delight. But it's the female birds of the same species that decide what is delightful. They go for the songs that move them. The brain of a female nightingale is chemically affected by the song of a really good male.

So you can listen to the blackbird and admire not just the fluting qualities of the whistling, but also the variety: the way the bird sets out on a voyage of musical exploration. The more successful this voyage, the more intimidating and attractive he is.

He will mix sweet notes with some more challenging ones, and he will throw in some passages that sound as if he was using an effects pedal, like a guitarist seeking a 'dirty' sound. He is very consciously a musician. That is to say, he doesn't sound, to human ears, like a creature driven blindly by pure biological ends. This is obviously speculation, but it certainly sounds to me as if the blackbird is lost in his own music. It sounds as if it's music first, biology second.

Birds sing in order to give delight. But it's the female birds of the same species that decide what is delightful

Song thrushes are also great garden songsters, and they're easy to pick out because they repeat. They throw in a phrase, repeat it two, three or even a few more times if they really like it, and then they move on to another. As they do so they take musical ideas from the world around them. They are not strict parroting mimics, but a sound with the right sort of pitch and tone will drive their musical ambitions and get them going.

If you are in a place with big mature trees, there are likely to be nuthatches around. One way of finding out if they're actually there is to listen to song thrushes, for they love to try out nuthatch variations. At my place, on a floodplain in the Norfolk Broads, the song thrushes often include the piping of oystercatchers in their repertoire. I knew that a greenshank visited a marsh in Suffolk because I heard a song thrush chanting like a greenshank.

Song thrushes also take on human-generated sounds. I have heard recordings of a song thrush mimicking a shepherd's whistle and a lawn-mower, and there was one I knew that imitated the warning signal of a reversing tractor. It sounds playful, and in a sense it is: but it's also the most serious thing a song thrush ever does in its life. That's because the bird with the widest repertoire has a major advantage when it comes to the task of becoming an ancestor. And that's a prize worth singing for. Singing your guts out.

Nightingales aren't night birds. They sing all day as well as all night. For a very brief few weeks in spring they arrive in the southern part of England and sing louder and longer and more flamboyantly than any other bird that breeds on our damp little island.

They're best heard at night because there's no competition. They belt out an extraordinary solo of intense whistles, powerful throbbing, reso-nant drumming and complex phrases of music, some of it as melodious as the sweetest blackbird, other bits as challenging as Schoenberg and Stockhausen. Even in the day they effortlessly surpass all those around them, drowning them out with a mixture of volume and extravagance, like a lead guitarist turning his volume up to eleven.

It's a song that's been hymned and praised and celebrated across the ages: Romeo and Juliet, John Clare, Coleridge, Shelley. People have said that the song is so wonderful they can hardly bear its continuation: it's too much, too intense, too demanding. Certainly it's not late-night easy listening.

One study showed that a male nightingale had a repertoire of 250 phrases that he had assembled from a vocabulary of 600 different sound units. All this is desperately demanding of the bird himself: he must have

the strength and stamina for that endless performance, and the power for its impossible volume: a nightingale in song can be heard from a mile off on a still night. He must also have the experience and the musicality to assemble the song in a manner that pleases some of the most demanding musical critics in creation: the female nightingales.

This bird is so remarkable and so distinctive that if you're in the right place at the right time, you will recognize a nightingale—even from a dense chorus of other avian singers—the instant the first note is uttered.

And as the bird continues almost to the point of exploding, so you will wonder why the song really needs to be quite this wonderful.

Marsh warblers can out-sing nightingales. At least in some respects. They aren't birds you come across often: only twenty to forty pairs breed in Britain every year. And they don't blow your head apart like nightingales. They are subtler, but a great deal more complex. They can sing for an hour without pause and belt out a song of such wit and inventiveness that it takes a trained ear to appreciate it. It's a bird you might pass by, but once you take the trouble to pause and listen, the bird's extraordinary brilliance comes to you. It's like some deceptive pieces of music: sounds simple but it really isn't.

A marsh warbler builds up an individual repertoire by listening to others of its kind—and also by listening to other birds. Research in a small population of marsh warblers showed that they imitated 212 different species of bird, and that each singer had an average repertoire of 76. They sing the songs and calls of birds that live and breed around them in the nesting area. They also sing like the birds they hear on their wintering grounds in Africa. And the sounds they pick up as they migrate between the two. If you have a good enough ear, you can hear a chaffinch's *finch* call melding into the call of a puffback from Africa, and a blackbird's song woven into that of the white helmetshrike, another African.

So we've established that birdsong is about territory. That is to say, it's about sex, procreation, food, protecting a female from marauders, raising young, becoming an ancestor. All good biological reasons. And that's enough to satisfy you if your mind is purely scientific. But it leaves a few things unexplained. You impoverish yourself if you accept only science, just as you impoverish yourself—perhaps more greatly—by ignoring science.

Birdsong is about sex, procreation, food, protecting a female from marauders, raising young

So consider this about marsh warblers. The song is specifically sung to repel rival males. But the song can also bring males together. You can sometimes find three or four of them in close proximity, all singing. They do so more quietly, more economically, than they do when singing territorially. They are singing for the sake of the song. For the music.

They will be swapping material, which will be useful to them, because the more variations you can sing, the more attractive you become. But the biological imperative seems remote from such jam sessions: it seems that the music is what matters. Or do you prefer to think that the birds are making long-term plans for the next mating season?

If you don't like the idea of birds as conscious planners, then you have to accept them as lovers of music for its own sake. Could any great musician give a great performance entirely for an ulterior motive? Entirely because at the end of the performance he was going to be paid? Surely not. Even the most money-grabbing of musicians gets lost in the music—music for its own sake. And I suspect that's true about the great avian singers that we find with such joy all over the planet.

And that's one of the things that birds and humans have in common. One more reason why birds are the interface between humans and the wild world.

4

*My place
or yours?*

W hen I was a boy I wondered why I didn't see more birds. It was a long time before I grasped the notion that you have to go to the right places. Birds are about place. No point in looking for woodpeckers in the desert and ostriches on the high seas. Not much point in looking for dunlins on Streatham Common either. I built a hide in the back garden and wondered why I only saw wood pigeons.

A bird *is* the place it lives in. It eats the place. It makes the sound of the place. The sighting of a bird gives you, the watching human, an instant understanding of the place you find yourself in. We often claim that a bird is the spirit or the embodiment of the place. Birds, more than any other living things, define the place they live in for the human observer.

An eagle, however briefly seen, seems to say everything we could possibly feel about a visit to the mountains. Many people visit Florida and find their experience summed up in visions of pelicans. A British back garden is the undisputed land of the robin. Birds unite the mobile human with the places we travel to and then they reward us all over again when we return home. A kestrel is the spirit of the motorway, and frequently, on journeys home from more exotic places than the M25, I've found the melancholy of leaving a fine place replaced by the consoling joys of home.

Birds sum up a location. That's why most nations have a national bird, and why there was so much interest in the recent project to select a national bird for Britain (see page 312). Birds pass on messages about the place in sound and colour, and all this is made more vivid by the fact that

previous page:
THE GANNET.
For caption, see page 89.

they could so easily leave. They have wings. They are highly mobile. It seems that they are here by choice, in the same way that we are.

There are about 10,000 species of bird in the world. So what's a species?

A species is (once you go above the level of the gene) the basic unit in which life makes more life. If you can mate and produce viable offspring—offspring that will be fertile and breed true in their turn—then the two of you belong to the same species as well as to opposite sexes. So, as we have observed, humans can't interbreed with chimpanzees even though we have nearly 99 per cent of our genetic material in common, and a willow warbler can't interbreed with a chiffchaff even though they are just as closely related and they look so similar that it's entirely possible for an expert to hold one in the hand and still make the wrong diagnosis. The best way to tell them apart—perhaps for the bird as well as the birdwatcher—is to listen: their songs are completely different, so presumably a female willow warbler isn't stirred when she hears a handsome male chiffchaff singing *chiff-chaff-chiff-chaff*, while a female chiffchaff feels her own heart sing in response.

So how do you get to be a species? It's a pretty massive question, but the birds explain it to us in a way that we can grasp intuitively. We don't need a scientific explanation: we just look at birds and it is sort of obvious.

Consider your bird table. Chances are you will get blue tits and great tits and coal tits: very similar birds, all closely related, though you can tell them apart at a glance. So there's a simultaneous togetherness and separateness going on here. They all share a common ancestor, but they have gone in different directions. They have chosen different careers, different ways of making a living. It's only on the bird table—a shared food resource—that they come into contact and clash with each other.

Great tits and blue tits forage for small invertebrates in the same tree, but great tits concentrate on the middle of the tree, while the smaller and lighter blue tits tend to hunt at the extreme end of the thinnest twigs, which are beyond the great tit's scope. So there, right in the same tree, we have different kinds of birds in different places. Coal tits are more likely to forage in a conifer, rooting out small scraps of life between the needles.

You won't get ducks hunting in the trees (though a few ducks nest in trees). They go for a different sort of place: one with water, obviously enough. They have webbed feet for more efficient swimming and mallards have beaks adapted for filtering small scraps of food they find below the surface of the water. You mostly find them at the edge, where the water is shallow. Swans will feed a little deeper, because they can reach further down, with longer necks and bigger bodies, and so they reach food unavailable to mallards. Out in the middle of the lake you will probably see neat black-and-white ducks: tufted ducks, or tufties. They can dive and swim underwater, so they can reach stuff that's deeper still. So all at once you have three species of bird. They all like roughly the same sort of place, but a different part of that place. They each favour a place-within-a-place.

Moorhens and coots are both black birds and they both have fresh water as their preferred place. They are related: they're both rails. But they can't possibly be confused. Coots live out in the middle in the deeper water, and can dive; moorhens like the tangled and weeded margins, and can clamber around them with immense dexterity. Similar birds: similar places: very different lifestyles.

And they trumpet their differences at each other, and incidentally at human watchers. Moorhens have a bright red beak and forehead shield; coots have the same parts picked out in an equally unmissable white. Thus (harking back to colour) they are using visual signs to recognize birds of their own species while emphatically not confusing them with their relations. They are separated by colour, also by call, and by their own preferred place-within-a-place.

left:
MOORHENS
like the tangled and
weeded margins,
and can clamber around
them with immense
dexterity.

In other words, birdwatching is also something that birds do. The decisive difference in the head colours of these two different species of rail is a signal both to their own species and to the potential confusion species. It's what birders call a 'field mark': something that allows the observer to make a confident diagnosis of the species the bird belongs to. Field marks matter to birds as well as birders. The ability to tell similar species apart is not something that humans impose on nature: it's the way nature works. And it's necessary when birds share a place but not a species. Each species needs to keep both to its own species and to its own preferred place-within-a-place.

There is always a point when birds with the same ancestor divide from each other, go their separate ways

There are two ways in which one bird (or anything else) becomes a different species from another. And they're both to do with place. Place defines a bird and what it is. Place sums up the global mystery of species: and species is the way that life on earth works.

The process of making a species is called, logically enough, speciation. There is always a point—even if you can't work it out with any exactitude—when birds with the same ancestor divide from each other, go their separate ways, find different ways of making a living and so cease to have anything to do with each other. They become two entirely separate populations, and so the process begins. The birds will increasingly diverge from each other: and over the extended timescales of evolution they will tend to become separate species.

Populations can start to differ from each other because they live in different places and, by mere geography, grow apart. Or they can discover different ways of making a living, even though they are still close together, and so they will start to separate and grow apart. So we have different-place speciation and we have same-place speciation. The scientific terms are 'sympatric speciation' and 'allopatric speciation': same country and different country. Readers are invited to supply their own Irish jokes.

Roughly 10,000 species of bird? Why roughly? Why can't they be sure? If a species is the basic unit of life, of inheritance, or life's continuation, then it should be pretty straightforward.

But it isn't. Take the crossbill.

A crossbill is a nice little finch. The males are a fine red colour when seen in a good light (at least by a human). It feeds on pine nuts: the pesto bird. It's called a crossbill because the bill is crossed at the tips: a perfect tool for opening up pinecones.

You can find them mostly on the eastern side of the country (seldom in huge numbers), where there are plenty of pines: I heard one in Greenwich Park during the equestrian events at the Olympic Games of 2012. The parrot crossbill and the two-barred crossbill are also recognized species, and they are clearly, if subtly, different.

And then there were four. In 1997, ornithologists decided that there was a fourth species: the Scottish crossbill. It looks just like the Greenwich Park crossbill—more usually called the common or red crossbill—but it makes radically different sounds. The song is different, and the flight call and the excitement call are very different. Further research has shown that the Scottish crossbill is 'reproductively isolated'

below:
THE CROSSBILL
feeds on pine nuts—
the pesto bird.

from other species of crossbill. There are common crossbills in Scotland, there are also Scottish crossbills—and the two species don't mix. They live in the same place, but they don't interbreed. Sympatric speciation.

This is Britain's only 'endemic'—a bird found only in one country. As such, it would have been an appropriate choice as Britain's national bird—after all, no one else has got one. If you hear a Scottish crossbill you know you can be nowhere else but Britain. But England and Wales were never going to accept that.

So the question of species is not as straightforward as it seems. The problem—unsettling for physicists—is that biological truths are a little bit fuzzy round the edges. Scientists change their minds on the question of when a species is not a species. Sometimes birds that are considered to be separate species will be reconsidered, and reclassified as one and the same thing; sometimes the opposite process will take place. Some scientists will have a taste for making more species, others for fewer. In the jargon they are referred to as lumpers and splitters.

The birdwatchers who value their lists of birds are always delighted by new cases of splitting. In 1997 those birders who had seen a crossbill and what was considered a Scottish race of common crossbills suddenly found they had seen two species: (common) crossbill and Scottish crossbill. So their bird list went up by one. It's called an 'armchair tick' and it brings great joy to those who care.

To the rest of us, such a decision is one of the many complex and often unfathomable mysteries of birds and place.

There are fourteen different species of kestrels in the world: small falcons, some of them more closely related than others. They have spread out across much of the world and adapted to meet the challenges of different environments. Different species of kestrel can act differently and look differently. However subtly.

Kestrels reached the Indian Ocean islands of Réunion and Mauritius. Being island birds, by definition isolated from others of their kind, they went their own way in terms of evolution. It is accepted that they became two different species, different again from the ancestor they shared: and thus we have the Réunion kestrel and the Mauritius kestrel. The two populations were reproductively isolated by all that ocean: allopatric speciation. The Mauritius kestrels don't have the sharp-pointed scissor-wings we normally associate with kestrels: they are rounded, more like a sparrowhawk. Sparrowhawks hunt in forests and need to be highly man-oeuvrable at speed. The Mauritius kestrels have adopted a similar life-style and so, though they are not closely related to sparrowhawks, they have come up with the same solution to the same problem.

In this way, islands often develop idiosyncratic species of their own; hotspots of endemism. The (relatively) nearby island of Sri Lanka has twenty-six endemics and seven more proposed: that is to say at least twenty-six species of bird found nowhere else in the world, and maybe thirty-three. Which makes it also a hotspot for list-conscious birders, and a fascinating place for anyone with a sense of wonder.

Island populations are by definition small and also by definition vulnerable. There are no small disasters for an island population: they can't recruit from elsewhere and build up again. So it was that the Réunion kestrel became extinct in the seventeenth century, and in 1974 the Mauritius kestrel was down to four individuals. No one's a hundred per cent certain why we lost the Réunion kestrel, though persecution and the introduction of alien species certainly helped. The Mauritius kestrel's tumultuous decline came about because the reckless use of the pesticide DDT had affected the food chain and effectively stopped the kestrels from breeding; more on that problem later. At the same time, introduced mammals—cats, mongooses, the monkey called the

The Mauritius kestrels don't have the sharp-pointed scissor-wings we normally associate with kestrels: they are rounded, more like a sparrowhawk

crab-eating macaque—pinched eggs and preyed on young birds.

This is when conservation efforts kicked in, led by what is now the Durrell Wildlife Conservation Trust. Eggs were taken from wild birds and hatched, and the young birds were reared in captivity. At the same time the wild birds were given food, to give them energy to lay again. By 1984 the population was up to fifty individuals, and by the early 1990s the population was self-sustaining, and captive rearing was scaled back. There are now several hundred birds, even though the population has dipped from a 2005 high of 800 adult birds.

This is a story that shows, among other things, how powerful the idea of species is to us humans. If scientists had told us that the Mauritius kestrel was a mere sub-species of the kestrel that haunts the M25, it's unlikely that we would have been so determined to keep it from extinction. The fact that it is something special, something unique—able to breed only with its own kind—has a powerful hold over the way we see the world, and over the way the world actually works.

And it is all about place.

The most dramatic way to understand the link between bird and place is to take a boat out to sea. Travel for about a mile—just far enough to make the shore look disconcertingly little—and it's like going through the wardrobe into a different world. The fact that you are in a new place is made clear by the birds.

We talk about seagulls, but that's the wrong name for them. Better to call them shoregulls. They operate best at the margins: they are edge-workers who mostly inhabit the zone where land and sea come together. They forage on land, they fish in coastal waters, they steal chips from humans and follow fishing boats for a free meal. But the deeper you get into the sea, the fewer gulls you see. Seagulls are not, in the main, seabirds.

When I looked at bird books as a boy I thought I would never see a gannet. I thought they were impossibly rare, occasionally seen by a lucky and particularly intrepid birder. But once you make that trip a mile from shore, you start to find them, and in the summer months you are likely to find many. These dramatic birds—the classic design of white with black wing tips (remember that melanin; see page 43) and a span of 2 metres (6 ft plus)—are as much the spirit of these waters as the robin is of the back garden.

And as you look, you begin to see another kind of bird, one you almost never see from the shore: dark wheeling shapes that cruise close to the water, seeming to play chicken with the white-topped surf, daring the waves to reach up and knock them from the air. These are shearwaters: an unusually apt name, for as they fly they frequently flip into a sharp turn, standing on a single wing tip that seems to slice the ocean below them as they wheel and then regain the horizontal. I remember an hour-long passage of Manx shearwaters that I saw from a boat off Newquay in Cornwall. We were looking for dolphins and basking sharks at the time, but that's the wild world for you—you never know what's going to turn up. And sometimes the consolation prize is better than the thing you were looking for.

Another boat. This time out of Stonehaven in north-east Scotland. A day when the mist allowed about a hundred yards of visibility. Sounds like a bad day for a seawatch, but it turned out to be an enchanted afternoon. It was as if this boat was at the centre of an upturned bowl of clear light, the birds entering and departing this bowl as we cruised along the forbidding rocky coast.

Birds appeared, birds disappeared. You had them for, say, twenty seconds, and then they were gone forever. And it was a parade of sea-birds. Real seabirds, not the shorebirds, not the half-and-halfers like the gulls. A different place, so naturally—*naturally*—there were different birds. Gannets, of course, and shearwaters. Also skuas, those dark, sinister relatives of the gulls, looming out of the mist like Dracula scenting fresh blood. And, of course, the auks: guillemots, razorbills and puffins.

Auks can fly. Just. It's not what they're best at, but it's how they get from their clifftop nesting places to the fishing grounds and back. They fly in a great frantic panicky-looking whir of wings; they look as if they are powered by wind-up elastic and at any moment the stretch in the rubber band is going to run out. In ones and twos and in little flocks they fizzed into our small bowl of milky light and out again.

Once they are on the water rather than over it you can see their real confidence in these conditions. They are birds of the open ocean, and they sit on its surface, buoyant as corks, serenely unworried by the hefty waves. They don't really care for land at all. They just need a few square inches of the stuff—preferably inaccessible to land predators—where they can lay eggs and rear chicks. Once the annual frenzies of breeding are over they go back to the sea, and I've no doubt they do so with relief, for this is their real home. They sit on the surface, comfortable in the big, heaving seas, and then they surface-dive and swim beneath it, powering themselves along below the waves. Here their short, stubby wings come into their own: they fly underwater, propelling themselves with these wings. The glorious, clamorous, stinking seabird colonies are home for just a few brief weeks. Their real place is the sea.

You don't find auks in the seas at the other end of the world. You'd have thought that these southern seas would be perfect for a bird that swims underwater and catches fish, and you'd be right, but there are no auks there. The same niche is filled by penguins. Different place, different birds.

There are eighteen to twenty species of penguin, depending on which scientists you agree with. They are found in Antarctica, Chile, Argentina, South Africa, Australia and New Zealand, and there's also a species, the Galápagos penguin, that lives where its name suggests. The Galápagos Islands are near the equator, but these penguins feed in the cold, rich waters of the Humboldt Current.

The group ranges in size from the emperor penguin, which stands more than 1.1 metres (3.5 ft) tall, to the little blue or fairy penguin, no more than 40 cm (16 in) tall. Their place is in the oceans below the equator, just as the auks have the oceans to the north. Like

the auks, penguins need to come on land to breed. They don't fly, so they need shores accessible to their clumsy waddling.

There was a time when the northern seas had a bird that looked just like a penguin, and lived pretty much like a penguin. This was the great auk: another flightless fisher. Auks and penguins have the same colour scheme: black above, pale below. So they look like the sea if you are looking down on them from above, and like the sky if you are looking up at them from below: helpful colouring both for catching food and evading predators.

The great auk was pretty impressive and bigger than some species of penguin, standing up to 75 cm (33 in). They went extinct in the nineteenth century, hammered by humans who used them for food and for bait. They also used their down for insulation. So the North Atlantic was the right place for large flightless birds that could fly beneath the waves: but alas, though we still have the place, we no longer have the birds. More on great auks later (see page 228).

I can hear firecrest and capercaillie: I must be in the Caledonian forest

Birds establish place. That's a truth known by film-makers. If they want to tell you that the action takes place in an English garden, they'll give you a bit of blackbird song: a subliminal clue that puts the viewer almost literally in the picture. If you move to the shore of a lake, chances are you'll hear the quack of a duck, usually a female mallard. If you're at the seaside, you'll hear gulls, usually the display calls of herring gulls—the sound you hear at the opening of *Desert Island Discs*. If you're on a blasted heath and you want a bit of sinister atmosphere, you'll hear the cawing of carrion crows. And if you're up in the mountains, within ten seconds you'll hear the call of an eagle. You don't often hear such a call when you're in the mountains for real, but you always hear it as soon as the action of the film shifts towards the peaks.

It follows, then, that you could transport a half-decent birder anywhere in Britain blindfold, and with a bit of luck and a good deal more knowledge, that birder should be able to tell you where you are. I can hear wood warbler and pied flycatcher: I must be in the wet woods of Wales. I can hear firecrest and capercaillie: I must be in the Caledonian forest. I can hear reed warbler, sedge warbler and bittern: I'm in the wetlands of eastern England. I can hear Scottish crossbill: I wonder what country I'm in now...

A field guide is a book that helps you to identify birds. I've just had a quick count of the field guides on my shelves: there are about forty, though I have a few more that I've put back on the wrong shelves and can't lay my hands on right now. I haven't collected them in any sense of hoarding and enjoying possession for its own sake; I have other vices. It's just that whenever I travel somewhere I like to know what birds I'm looking at. And it's no good taking your British or your European bird book to China or Brazil.

Lord, what glorious, intimidating volumes they are: pages and pages of birds often completely unknown to you—and mostly to me. In unfamiliar places—for me that's Australia and South America—it's often hard to work out which family the bird belongs to, never mind what species.

You don't know what page you should be looking at, or even which bit of the book.

But as you begin to learn a place you begin to learn the birds. In time the daunting pages of the African field guides have become almost as familiar to me as the pages of British birds in the volumes I left at home, so that when I was in mopane* woodland I knew that the black-and-white flicker I saw would be Arnot's chat, and at a certain bend of the Luangwa River there would almost certainly be Böhm's spinetails in the air. I grew to understand the Luangwa Valley by learning to understand the birds. A place and its birds are in a sense one and the same thing.

So a really good birder—a birder with universal knowledge—could be drugged and transported anywhere in the world and, once revived, would be able to tell you the name of the country and what part of that country you were now in, and whether you were in wetlands or mountains or forests.

If you hear a strange wailing cry, haunting and a trifle unhinged, and you know you're in Britain, there are only two places you could be. Well, make that two and a half. For this is the voice of the stone curlew: a mad, goggle-eyed bird that loves darkness and the bare, stark, arid country. Stone because it makes its living on uninviting stony ground; curlew because it sounds a bit like one, though they're not closely related.

The call of a stone curlew means you are either in the Brecks of Norfolk or on Salisbury Plain, though it's just possible that you're on the Suffolk Sandlings. These are the only places in Britain that provide

I grew to understand the Luangwa Valley by learning to understand the birds. A place and its birds are in a sense one and the same thing

* A tree that grows in hot, dry, low-lying areas of Namibia, Zimbabwe, Botswana, Zambia and Mozambique.

the dry country these birds live on. That means open heathland that is fiercely grazed by rabbits, or on spring-sown crops.

This at once presents a problem if you want to keep stone curlews, for they have declined sharply with the spread and the intensification of agriculture. You have to look after the heaths and make sure that they are fierce, short and dry enough for the stone curlews; and you have to persuade farmers that stone curlews are a good thing.

It follows, then, that if you want to look after birds, you have to look after places.

The red-billed curassow is a rather comic-looking beast, big and scruffy with an air of slight bewilderment. It's big and meaty, and the males have a rather clownish Red-Nose-Day beak. They are birds of the *Mata Atlântica*, the Atlantic Forest, which is perhaps the most comprehensively destroyed habitat on earth. It's on the Atlantic coast of South America, mostly in Brazil: and that's where the rush to develop the rainforest has been most destructive. There's only 7 per cent of it left.

It follows that if you want red-billed curassows you have to look after the Atlantic forests we have left, and to try to extend them. If you want the bird, you must have the place. That's the rule in the *Mata Atlântica*, in the Antarctic, in the African savannahs, in the Brecks of Norfolk and in your back garden.

So I went to visit REGUA in Brazil, about a ninety-minute drive from Rio. Full name: Reserva Ecológica de Guapiaçu. Here Nicholas and Raquel Loke run a project that safeguards and extends the Atlantic

Forest. There are approximately 7,300 hectares (18,000 acres) of good forest there, along with a captive-breeding and re-release programme for the red-billed curassow. I am an ambassador for the World Land Trust, which has given great support to REGUA since the project began in 1981.

I threw a leg over one of the tough little horses that abound in the wilder parts of Brazil and rode uphill for half a day into the least

right:
CRESTED CURASSOWS are large ground-dwelling, rather comic-looking birds of the humid forests of northern South America.

accessible part of REGUA, savouring this sense of place: of trees, moisture, growth and life. Then we came down again with my boots higher than the horse's ears. This is what place means. It's no good breeding red-billed curassows if you haven't got any *Mata Atlântica* to release them into.

Birds teach us that place matters. It matters to them: but it also matters to humans on all kinds of different levels. It's good for our souls to know that there are rich, wet tropical forests out there. It's also been shown to be essential for maintaining our climate and the health of the entire planet. We need to safeguard the future of the red-billed curassow—out of pure self-interest.

5

Dancing
to the music
of time

They come bearing the gift of summer in their beaks. Who was the misery guts who said that one swallow doesn't make a summer? Better to salute the pioneer bird and wish him well and hope that the birds that inevitably follow are as bold and as capable of spreading joy.

Birds are about time as well as place. Like us all they have their being in the dimension of time. They can tell us about the time of day and about the time of year, and about the passing of time across the millennia.

The first swallow of the year is a moment of annual wonder. Swallows appear like a miracle, as if a fairy godmother had suddenly willed them into existence: and there they are, flying in their usual curves, circles, spirals and arabesques as if they've never been away: both surprising and inevitable, like all the best drama.

The first swallow is the striking of the clock of the year, telling you that it's half past spring and summer will be along in a minute. They are positive proof that the dragon of winter has been slain, that we can now walk more freely, prepare to cast a clout and plan jollier, lazier and more vigorous things than seemed even possible a few weeks back.

Swallows are birds of good luck, happiness, hope. They are a traditional part of the tattoo artist's repertoire: a swallow embodies a sailor's hope of seeing land once again. And they're birds of hope for us all because they're associated with fine weather—after all, they need fine weather or they die. That's because they make their living from aerial

previous page:
THE SWALLOW.
For caption, see page 101.

plankton: the invertebrates that can be found in the air.

Such creatures can only survive in warm weather. In Britain you seldom encounter insects outdoors during the colder months. But then they hatch out in numbers and swallows come swooping in to catch a bonanza that would otherwise be unexploited.

Swallows are specialists and the species of swallow that we know in Britain is committed to long journeys to chase their specialization. For the bird it's a survival plan and one that has been wonderfully effective across the millennia. For humans it's a rekindling of hope.

above:
A SWALLOW embodies a sailor's hope of seeing land once again.

Gilbert White was one of the great naturalists of all time. He lived and worked only in Selborne, a small local patch in Hampshire, and he understood it holistically long before the word had been invented. As such, he was the founder of the science of ecology, though he never knew the term. He studied live animals in the wild, rather than dead skins, so he was also the founder of the study of animal behaviour, or ethology. He was fascinated by the way nature operates in time. The passage of the seasons enthralled him. Thus he also invented the science of phenology: the study of the seasons: wild time.

Swallows enthralled him more than any other species. Where did they go in the winter? No one knew: so the magic of their reappearance was still more vivid, being apparently inexplicable. He worried away at this question throughout the course of his life, and throughout the course of his life's work, *The Natural History of Selborne*. And, almost reluctantly, he sides with the theory that swallows hibernate over the winter, that they hide away in the mud beneath the surface of ponds and emerge when the weather is fine. The fact that the first swallows of the year tend to appear over water seemed suggestive.

But the truth is, of course, more wonderful. In the wild world, that is usually the case. Swallows are long-haul commuters, flying between northern Europe and sub-Saharan Africa twice a year, feeding on aerial scraps as they go and bringing joy wherever they are sighted. They are often first seen over water because that's a good place to hunt for insects.

Thus swallows link modern humans to the turning of the natural year, reuniting us with our wild birthright.

A swanfall is one of those routine miracles that the wild world throws at us, and it's as wonderful a thing as I've ever seen. First the lake was open and pretty empty: within the hour you could see nothing but swans. It was like watching Bank Station in the City of London in the morning: a place that is at first sparsely populated turns into the busiest place in the world before your eyes: not gradually but all at once.

Not harassed and resentful commuters, though. Swans. Swans making this landfall—or, to be strictly accurate, waterfall—with joy and relief after a long, hard journey. Swans coming home, and doing so in

Bewick's swans arrive at Slimbridge family by family; the first thing they do is sort out the vital question of who defers to whom

vast numbers. These were not the swans of summer, the stealers of bread from duck-feeding children, the harassers of anglers, the posers of the Avon. Not mute swans. Not an everyday part of the British waterscape. These were Bewick's. Wild swans.

Bewick's swans are named for the great engraver and author Thomas Bewick (see page 265). They breed in the Arctic, around the Kara Sea in Russia. They come here for the winter: for them, the bitterest days of our January are balmy. They make a journey of 2,500 miles (4,000 km) via the White Sea, the southern Baltic, the Elbe estuary and the Netherlands before arriving tumultuously at a few sites in Britain.

One of these is Slimbridge, headquarters of the Wildfowl and Wetlands Trust, which was founded by the great naturalist Sir Peter Scott in 1946. Scott was, among many other things, a painter with a painter's eye for detail and colour. And he noticed that the black and yellow markings on the birds' beaks were variable. As he took pictorial notes of this, he realized that each beak was as unique as a fingerprint. Thus he began a study of the birds as individuals, which survived his death in 1989 and continues to this day.

The Bewick's swans arrive at Slimbridge not one by one but family by family, and the first thing they do is sort out the vital question of who defers to whom. They establish precedence gang-handed. One family group will swim at another with necks swaying, uttering their quasi-musical bugling call in unison. A group that feels mastered will back away. If they feel otherwise, they will steam forwards with their own counter-bugling response.

It's mostly a matter of numbers: a group of four will be mastered by a

group of five, and be in turn mastered by a group of six. It follows that a pair that manages to raise four young in a year will have better access to resources on the wintering grounds than a pair that only raised a couple. A group that takes casualties on the migration run will find their chances of surviving the winter compromised.

The swanfall makes it very clear that winter is now close at hand. These white arrowhead formations in the sky that make their way across the country, bugling as they pass, are bringing with them darker and shorter days, colder and more hostile weather. To humans they are a powerful symbol of life's determination to continue, despite the inevitability of more difficult conditions. They are birds of strength and purpose. They add a beauty and a sense of resolution to days we tend to endure more than enjoy. If the swans can cope, then so must we. They are birds of the darker times of the year: but they bring with them a certain light and hope as they come.

Winter is much shorter than we think. All we need to understand that potent fact is a pair of ears. By the end of January it already seems that winter is taking up far more than its fair share of the year, but if you can tune in to the birds you learn to hurry the year along a bit. You soon understand that the passing of the seasons is a more rapid thing than we fear on a murky journey home in the blackness of the late afternoon. A person with tuned-in ears—a tuned-in mind—can start counting down the days of winter several weeks earlier than a person who assesses the time of year by the number of garments that can be cast aside.

facing page:
A GROUP OF SWANS.
In the foreground
there is a whooper swan,
with a mute swan behind,
and Bewick's swans
behind and to the left.

A sunny day in early February, often a frosty see-your-breath morning: and the air is filled with a loud, confident, even defiant, two-syllable call. This is a great tit singing in the first days of spring, as he reckons it, defying the winter to do its worst: teacher-teacher-teacher! The song is routinely compared to a squeaky bicycle pump.

Great tits have adopted the strategy of earliness and it has become their great advantage. They are among the first species to set up and establish a territory, woo a female and get down to the business of breeding as early in the year as possible. We humans naturally tend to see this in a less business-like way, as something less fuelled by Darwinian imperatives. To us, the teacher-teacher song sounds like the first really solid blow against the forces of winter. It brings great good cheer to us now, as soon as we have learned this simple song, so think what it meant to our ancestors, who lived their lives season by season, infinitely closer to the bone than we can imagine in the twenty-first century.

Great tits have adopted the strategy of earliness and it has become their great advantage

The great tit begins it all. What follows is not a great unison chorus but a slow and deliberate build-up. Winter is not vanquished all at once; it dies the death of a thousand songs. Wren and dunnock both like to celebrate warm days throughout winter with a quick burst of song, but as they get more hours of daylight, so they respond with more and more singing. Song thrush then cut in with their repeating songs. Chaffinches come in with the second or third wave of singers. Greenfinches are usually a little behind them, at first singing almost reluctantly, as if they mistrusted the year, but then they acquire confidence in the spring and let rip.

It's the blackbird that signals the completion of spring part one, joining the singers with that lovely laid-back whistle: the great flautists of suburbia. And then it's time for part two.

The first migrants tend to hit southern Britain in mid- to late March and move north. The migrants are phase two, but phase one and phase two always overlap. The first migrants to arrive come from southern Europe and North Africa. There's a trio to look out for: sand martin, the slightly subfusc relative of the swallow; wheatear, which mostly pass through rapidly, spelling out their presence with a gleaming white bum as they fly away (wheatear is in fact a contraction of 'white arse'); and more or less at the same moment, or even a little earlier, the chiffchaff strikes up. This is another two-syllable song: sweeter and less strident than the great tit's.

Not long after this the blackcap joins in: a glorious song that some people prefer to the nightingale's. It's even called the northern nightingale, because it penetrates much deeper into our country than the nightingale, which tends to hug the south-east.

The long-haul travellers come in last of all, after a journey of thousands of miles. Willow warblers are tiny things and yet they cross the Sahara to be with us for the summer, many of them travelling as far as northern Scotland. Their song is not over-complicated—a sweet lisping descent down the scale—but it has always been a special one for me. It seems to signal the end of the beginning, that winter has been truly vanquished and now the great life-giving season of the year can begin in earnest.

Birdwatchers tend to become butterfly-watchers in the high summer. Summer is the time for butterflies, creatures of the sun and flowers, the males interested only in drink and sex—in other words, nectar and female butterflies. The females have more to do, for they must dispose of the eggs on the appropriate food plant. This silent wash of flickering colour across the countryside is an annual delight. It's not the time to be too single-minded about your favourite kind of flying creature.

Butterflies tend to peak at the right time for birders because summer is the quiet time for birds. Quite literally, in many cases. Once songbirds have started raising young they do very little singing. That part of the job has been done: a territory has been established and a female enticed into it. The hard job of parental care now dominates everything. This is what it's all for: the hanging tough throughout the winter, the establishing of territory, the singing, the wooing, the nest-building. This is the time that really counts: make or break for the genes. So it's one of those wild paradoxes: the most hectic time of all for the birds is, for the human observer, a time of quiet acceptance.

In the summer birds are less audible and less visible. They are not moving about much, for there is no reason to do so. To a human it feels like a time of peace and repletion, of lazy revelling in achievement. Our response to wild events is often at odds with reality. An owl's hoot in a graveyard is scary for a human, but reassuring to an owl; a bird's song is relaxing to a human but a frenzy of emotions for the singer and for the avian listener; a swallow crossing a field is an infinitely peaceful sight for us humans, but each jink in the flight is a little death of some flying creature.

An owl's hoot
in a graveyard is
scary for a human,
but reassuring
to an owl

So we can savour this contradiction and enjoy the apparent calm of summer... while keeping our eyes open for fritillaries and hoping that this will be a painted lady year or a clouded yellow year—or even both.

There are two kinds of birders. There is the May person, and there is the October man. They're almost all men, the October lot. In May you take in the spectacle of the high spring and the great achievement of the wild world. It's kind of a set piece. A human observer pretty well knows what's coming and what's coming after that. The details vary, but the broad pattern is always the same.

But in October birds are on the move and anything can happen. The serious spring-and-summer business of breeding is all over. Now it's time to move on to the next phase. And birds, being equipped with wings, are immensely keen on movement. So the migrants get ready to migrate, gather in pre-migration flocks and in the end they get their act together and go.

You can watch the swallows on the telegraph wires: well done, little one! You flew all the way from the barn to the wire. And now for another flight. Cape Town.

Birds depart, birds arrive. Resident British birds are also on the move. Many species give up being half a pair and become members of a flock. Many individuals spend their winters in quite a different place from the breeding grounds: curlews fly down from the high moors and come down to the river estuaries; great crested grebes often move from rivers and lakes to more open waters and the sea. October is a time of urgent and dramatic transitions, and in that time all kinds of birds can turn up in all kinds of unexpected places.

For the October men (and, yes, there are a few women) this is the time for hunting. It's a process that for the best of them requires all the hard-won skills in identification and in knowing the right places to look. For many birdwatchers, this means an annual trip to the Scilly Isles. Here, on this outpost of England, extraordinary species touch down every year: lost, windblown migrants that missed their step. So if you are anxious to see Fea's petrel, sociable plover, yellow-browed warbler, Blyth's reed warbler, buff-breasted sandpiper and red-flanked bluetail, Scilly is waiting for you.

It's not my thing, but there's no call for me to get snobby about it, unless I happen to be in the mood for teasing twitchers. (A twitcher, you should understand, is a specific kind of birder: a dedicated seeker of rarities. We're all birders, but only some of us are twitchers. More on that later.)

It is the way that birds have their meaning in time as well as space that drives this human hunger for rarity-spotting. We love to hunt for something special.

Or you could go to Zambia, which is my preferred place for the annual avian Oktoberfest. But I insist on the far end of October, moving on into November. This is another of those dramatic collisions of the seasons and it's marked by the movements of birds and by the alterations in their behaviour.

In Zambia—as in much of Africa—the seasons don't blend subtly into one another as they do in Britain. The year turns on a sixpence. It changes with a bang. Quite literally. The rains come, the thunder speaks, lightning cuts the sky in half and rain falls on the thirsty land as if from a giant inverted bucket. I have seen the great thunderheads steaming across the sky, apparently being towed by European swifts: birds who have abandoned Europe to surge into southern Africa, surfing on the weather fronts and screaming at the tops of their voices.

Overnight everything changes. Suddenly the trees are ringing with a call you never hear during the dry season. The woodland kingfisher is back, and waking the wet season in the manner of an alarm clock. The colour values change, green comes back into fashion and the papyrus swamps are livid and vivid with new growth. And among these leaves you can see startling detonations of red. A few weeks ago this was just another LBJ: a little brown job impossible to distinguish from many others without considerable birding skills. Now in the season of plenty he has turned the most startling shade of red. This is the red bishop, a bird that chooses a rainstorm as his cue to burst into flame.

The paradise whydah performs a similar trick. In the dry season the male is another inconspicuous bird, but once the rain comes he develops a huge and extravagant tail, twice as long as his own body. It's both a handicap and a powerful turn-on at the same time. The male is telling the world, and especially the (to a human eye) unexciting-looking female paradise whydahs, that if he can thrive and prosper even with this

preposterous tail, he really must be one hell of a bird.

Thus the length and breadth of the Luangwa Valley is transformed, and the birds tell us how and why, and they inform us by filling our eyes and our ears with the thrilling urgency of the new season. Different species of cuckoos fill the air with simple, far-carrying cries. The yellow-billed storks return en masse to their great communal breeding ground and they set up their nests with deafening bill-clattering conversations: communication by means of percussion. It passes among stork-kind as a voice. And out in the soft and newly refreshed marshes, the tall and stately crowned cranes nest on the perilous ground that won't take the weight of anything more substantial than their lovely selves.

And as all this happens the African birds are joined by birds more familiar to us. The European swallows fly in and make their eternal curves over the wooded savannahs with the same swooping insouciance as they flew over the meadows and waterways of Europe. You can hear the garden warbler from an acacia tree, and—always a favourite moment of mine, this—you can hear the sweet and lovely song of the willow warbler.

'They're not European birds at all,' an African birding friend said to me. 'They're African birds.

They just happen to breed in Europe, that's all.'

But it's not so much about place as about time. The changing times of the year can be understood in the movements and in the behaviour of the birds. That's true all through the year and all over the world.

Birds also tell us about the time of day, though in the warmer months this can be a little elusive to humans. Unlike us, they adapt their schedules to the time of year: to the hours of daylight, to the food available, the tasks ahead, to the number of beaks they have to feed. For most species, day starts at daybreak. By the time humans are up and about in May and June, birds are already taking it easy, in the middle of a mid-morning pause.

There is a disconnect between us and the wild world here. We tend to take our walks just after lunch, and often that is when birds are at the quietest point of their day; they will tend to get busier towards late afternoon. The warmest part of the day is often the least suitable for birds.

But dawn is essential. There is urgency for food. Flying is so demanding in terms of energy that birds need to refuel as early as possible. It's literally a life-and-death matter. That's especially true when there are young to feed, for an unfledged chick is only ever hours from starvation. Here the short northern hemisphere nights are a major advantage to a bird—so many more daylight hours to feed in. That's one of the reasons why so many birds migrate here to breed: not just the availability of food, but the generous hours in which to find it.

In the winter, when a day is just a short break in a season of darkness, it is essential for many birds to top up at dawn. That is the real emergency of the day. Winter is a time for surviving, for trying to get through the months of deprivation in order to breed again.

In evening the emphasis shifts from food to safety: twin ambitions for survival but their relative importance shifts at different times of day. A bird must often take risks to find food, because the alternative is death. But the evening is all about finding a safe place to roost. When you take a walk on a winter afternoon, you will see birds finding a place of refuge at the top of a tall tree. Often they do this surprisingly early; at two in the afternoon, say, they will be settling down for a sixteen-hour night, in which they shut down and save energy. If you have had a successful day of foraging, there's no need to take any more risks. Quit while you're ahead, and hope to find a good food source at first light.

This evening roost can sometimes be a dramatic business. Some species of birds like to roost in large parties, as we have seen with starlings. The more of you there are, the greater the odds against being taken by a predator. It's always more likely to be one of your neighbours that gets got: that's the heartless imperative behind group solidarity of all kinds.

Often the roosting will be accompanied by a lot of fuss and noise and squabbles for the best places, with new birds coming in, some birds being displaced and coming in again, to displace some other luckless animal on the better and safer perches.

Sometimes this roosting cacophony will be accompanied by more elaborate behaviour, like a starling murmuration. There's a great crow

A bird must often take risks to find food, because the alternative is death

roost, used by carrion crows, rooks and jackdaws, not far from me, and at dusk in winter the sky goes black long before the light has gone as the crows wheel and turn in the air, cawing and jacking to each other, celebrating togetherness. Flocking is a survival strategy, but this loud daily celebration of togetherness makes the whole business more attractive, more meaningful, and therefore more effective for every bird involved.

Dusk is the moment for some birds to come into action. Find the right kind of rough, open countryside and wait there an hour or so before dark, and you may well get lucky. These periods in between day and night are barn-owl times, when you can see that subtle pale shape working soundlessly along the line of a hedge or a ditch.

They are birds of low light rather than pitch black. They are crepuscular, birds that work the boundaries of the field and of the day. I've seen barn owls respond to an approaching thunderstorm; the day blackens and the owl, roused from daytime slumber, tries his luck in the false dusk that has abruptly descended.

The heart of the night is the preferred time of the tawny owl. We humans have always seen light as good and dark as bad, so the hoot of a tawny owl is for British people the most sinister sound in nature. But these owls love the dark because it's an effective strategy. Tawnies make the dark work for them.

They can see very well, and they can hear very well. That means they can hunt very effectively in their chosen wooded places. But they gain their edge from local knowledge. A tawny knows his woodland as well as

above:
A TAWNY OWL
knows his woodland
as well as you know
your house.

you know your house, and they can find their way in the dark as well as you can. You don't need to turn on the light for a midnight pee; you know the place too well for that. You and the tawny are both capable of interpreting meagre information in a meaningful way: that small gleam is a metal door handle, that darker shadow is a step down, this pale line is a favourite hunting perch, that dark patch is a favourite killing ground.

In this way tawnies make the darkness work for them: a black shadow descending from a dark sky. But if you blunder into the wood and panic them, a tawny is quite capable of flying straight into a tree and breaking his neck. A tawny's senses are very good, but they're not miraculous. They need to be allied with knowledge and intelligence. That's how tawnies make the night work for them.

Birds can also tell you the time of the century. Birds, more than any other form of life, can show us when the times they are a-changin'. As always, their visibility and their audibility make hidden processes clear to human observers. Thus we have, for example, the inexorable march of Cetti's warbler and the little egret.

Twenty years ago both of these birds were exoticisms: birds to get excited about, birds to tell your friends about. They were birds that lived further south, adapted to warmer conditions, preferring milder winters.

But they started to be seen more often, and then more often, and now they are part of British life. Cetti's warblers first bred in Britain in 1972. Now there are 2,000 males singing annually—a song that's explosive and unmistakable; I can hear it in season from my desk. Little egrets first bred in Britain in 1996 at Poole, and within ten years the population was up to 500 breeding pairs and rising.

This shift is all about climate change; these birds aren't global warming sceptics. Instead of arguing about figures and interpretations and how many votes it would cost to do something about it, they followed the warming weather up north and have settled in very happily.

Birds are also very good at showing us the way we humans have changed the place for the better. They also reflect changing times and show that lazy cynicism need not be the default position. The most obvious example of this is the avocet, which became extinct as a British breeding bird around 1840 but came back during the Second World War. The land on the Suffolk coast that was flooded for national defence suited their needs and they recolonized, and they are now thriving. These days the population is up to 1,500 breeding pairs. Our estuaries are so inviting in winter that the population rises to 7,500 in the colder months. More on avocets later (see page 253).

There are plenty of other examples. Marsh harriers were down to a single breeding pair, at Minsmere in Suffolk, in 1971. Now they are up to 400 pairs breeding annually. Bitterns have also known local extinction in this country, and even after they came back they were struggling.

The places these birds needed—wet, fresh reed beds—were no longer being created by natural forces in a country that tends to maximize its resources of land for agriculture. As a result, bitterns were being squeezed out. But high-quality conservation work—creating the right habitats—has seen the bitterns return in decent numbers: a recent survey counted 80 male birds, establishing territories by means of the far-carrying and rather spooky sound they make that's called booming. The most obvious success has been red kites, which in Britain in the mid-twentieth century were down to a few pairs in Wales. In 1989 red kites were reintroduced to England (in the Chilterns) and Scotland, and they have spread spectacularly, up from nothing to 1,600 breeding pairs.

These birds show that we live in more enlightened times, times in which we humans can get things right, reversing the tides of destruction and working with the wild world to our mutual benefit.

People used to compete to hear the first cuckoo; now most people in Britain go through a year without hearing a cuckoo at all

Birds also show the exact opposite. Nothing demonstrates our own destructiveness as unequivocally as a bird. There are declining species of birds all around, most of them connected with farmland and affected by the shift in farming practices. Skylarks and lapwings are in sharp decline, and so are corn buntings: all birds associated with a benign British countryside.

The declines of some British birds are not just about local conditions. The migrants are also finding troubles on their travels, most especially the turtle doves, which are shot, frequently illegally. They have declined 95 per cent in recent years. Cuckoos are also in steep decline. People used to compete to hear the first cuckoo; now most people in Britain go through a year without hearing a cuckoo at all.

There was a time when people took canaries down the coal mines, because the canary was the first to be affected by the presence of toxic gases, and so acted as a warning to miners, who could either beat a retreat or put on respirators. Out in the world on the surface birds have been doing the same job year after year. A pity we don't pay them more attention. These are the rigs of the time, so the old song goes. Birds tell us about that, and do so with the utmost clarity.

6

The killing skies

Any little bird worth its salt will take on a bird of prey. It looks like a suicide mission: it's actually a diamond-hard survival technique. I was doing some outside chore when I heard the two-note call of a swallow. In easy times this is a cheery honey-I'm-home greeting as the bird enters the stable and makes a steep upward curve to its nest among the roof ties. Not this time. The same notes, but the tone was urgent, and much repeated. This wasn't a greeting; it was a call to arms.

I looked up at once, for the call is also a summons to human birdwatchers. At once I saw a hefty female sparrowhawk (females in most birds of prey are about a third bigger than males). And it was being harassed by the swallows from my stable block: zooming in at the bird like four jet fighters attacking a bomber.

This was doubly irritating for the sparrowhawk. It meant first that a hunt had been interrupted, and second that it was impossible to resume. There was no chance of taking anything by surprise with all that racket going on. Every small bird knows what these alarm and mobbing calls mean. I've got your number! On yer bike, hawky! And don't come back!

There's a passage in Anthony Powell's great twelve-volume novel sequence, *A Dance to the Music of Time*, in which, after experiencing the most extraordinary coincidence, the narrator observes: 'Life is full of internal dramas, instantaneous and sensational, played to an audience of one.' Wildlife does the same thing, and it does so over and over again. For the human observer, drama is almost the stock-in-trade of the wild

previous page:
SPARROWHAWK.
For caption, see page 137.

world. People talk about looking to the wild world for peace and calm and contemplativeness, but for any half-decent observer the wild world is also the source of unending dramas, played out with the utmost intensity, to the solo audience.

The drama of life and death in nature is utterly compelling, and we humans have pursued it across the centuries. Even today we shape vast acres of our countryside so that people—generally rich people—can witness it and participate in it.

It starts with birds of prey. Nothing in the wild world compels our attention like birds of prey. No group attracts us in the same way. We see them as glorious symbols of manhood; we see them as competitors and blood enemies. People love them with desperation; people want to see them wiped out.

Savour this for drama. Cornwall: uncompromising rock formations that hit the sea with startling abruptness. The restless sea, for once blue, but always tinged with white along the wave tips, around semi-submerged rocks, and where the water meets the land. The clifftops yellow with gorse and purple with heather. A bird riding the cliff's updraughts with insolent ease. Good enough if I stop it there.

But I don't. At some unseen signal the bird changes shape. Dramatically—of course, dramatically. It turns into an anchor; it turns into the Greek letter psi ψ. And it crashes. It disappears from view in the manner of a thunderbolt, only better. So, of course, it's a peregrine. One moment it's in the sky, the next it has vanished, beyond the limit of your vision, below the rim of the cliffs. You wait a while for it to reappear and climb again, but it does no such thing. Only one reason for that: death.

Peregrine. Peregrine falcon, a female from the size of her, making one of those 300-mph stoops, hitting a bird and killing it in an instant from pure impact shock. Would it have been more dramatic still to witness the moment it struck? Or was it better that the action took place, in the manner of a Greek tragedy, offstage?

Humans have always admired that moment of death, envied it, resented it and, above all, wanted it for ourselves. Throughout our history we humans have wanted to be that hawk. We have wanted to fly and we have wanted to use our flight with lethal purpose. We identify with the hawk, though we also identify with its quarry. We admire the killing stroke of a falcon, but at the same time something else in our make-up is rooting for the underdog. The underbird. Either way, or both ways at once, birds of prey have us entranced.

We want to be hawks. We want to soar, we want to fly with power and purpose, but above all we want to kill: or, at least, to play our own part in the great dramas of the sky.

And so we invented falconry. Perhaps the most ancient sport of all, for it is not the most efficient way of filling the larder for human feeders. Rather, it is something you can do if the larder is already full and you have time to spare to seek such things as beauty and drama. And, of course, personal prestige.

We admire the killing stroke of a falcon, but at the same time something else in our make-up is rooting for the underdog

All over the world people carry guns. Handguns, that is. Handguns are not very accurate and they have limited firepower, but killing people is only their secondary function. The principal function is to make their carrier feel just a little bit like God Almighty. Any gun makes you feel all-powerful; a special sort of gun—new or expensive or well loved or unusual or particularly well decorated or especially big—also speaks about your prestige and status.

It's also a fact of life that carrying a falcon on your wrist makes you feel just a little bit like God Almighty. I know; I've tried it myself. I've done a couple of falconry days. I've walked with a hawk on my fist and felt dizzy with delight at the way this bird perched there, its hooked bill a few inches from my left eye, its killer feet comfortably grasping my gauntlet. I let the hawk fly: I called it back with a bit of something raw and dead, and I saw the great beast swoop at me, so fast that at first I was inclined to flinch and duck.

I learned to control that reflex and hold still as a tree, my gloved hand an extending branch as the hawk approached, dropped low and then at the last second made a neat upwards curl to stop dead, losing all forward speed in the instant it alighted, a piece of flying skill that delighted me every time.

A hawk. A Harris's hawk to be precise, the hawk of choice for most falconers. They've been called 'the Labradors of falconry', and they have been selectively bred for their affinity for humankind. They are the species that adapts best to our demands; in the wild they are highly social and hunt co-operatively. They are the closest you can get to a pack-hunting animal in a wild hawk. It's a Harris's hawk that chases away the pigeons from Wimbledon during the tennis championships.

And for an hour this hawk was mine and I felt the thrill not just of ownership, but also of identification. It wasn't just a bird flying; it was my bird flying, and it was also my bird flying back to me. As if it knew me, as if it trusted me, as if—I dwelt in this fancy for the hour, at least—it loved me. Almost as if it *was* me.

Humans have gone in for falconry since the dawn of civilization.

The sport seems to have arisen from a combination of leisure and prestige. Civilization, if you like. It's generally agreed that falconry began 4,000 years ago in Mesopotamia, that insanely fertile patch of land between the Tigris and Euphrates rivers where civilization was also invented. Naturally that involved birds, because humans have forever felt this visceral connection with all avian life.

Falconry carried on as a mainstream activity for the wealthy and privileged more or less until the invention of the shotgun in the middle of the seventeenth century. At this point it is customary to quote some famous lines from The Book of St Albans of 1486, which assigns a different bird of prey to each social rank. I shall refrain here, because (a) you probably know it and (b) it's nonsense. The idea that an emperor goes hawking with a vulture is absurd: vultures don't hunt, and besides, the weight would have your arm falling off within the hour. Nor is an eagle—the other choice offered to an emperor in the book—a suitable bird. I once held a bald eagle on my wrist and it was a powerful experience, but not one that could ever last long. They're too damn heavy to carry for a day's sport.

The idea that an emperor goes hawking with a vulture is absurd: the weight would have your arm falling off within the hour

Falconry put deep roots into our culture. It was all about the intoxicating combination of fun and social rank. A lord with a trained peregrine on his wrist would indeed have felt that little bit like God Almighty—especially when he flew his bird at some really imposing target. A crane, say, or a heron.

There are expressions from falconry in twenty-first-century English, lurking in our language like fossils in living rock. Haggard, we say, meaning that someone is looking old, over-stressed, ill. A haggard is an adult hawk caught to be trained for falconry, so the bird looks different from a bird captured from the nest, and is much harder to train. If you are under a person's thumb you can't do anything of your own volition—like a hawk whose jesses (leg straps) are caught tight to the falconer's gauntlet, literally under his thumb.

When I did one of those falconry days, I remember another participant losing his hawk because he hadn't secured it properly. My own bird also attempted a flight on his own initiative and I vividly recall the spread of his wings and the tug on the jesses, but I—as a horseman more used to handling animals with a bit of wildness in them—was able to restrain him. I kept him under my thumb.

A *lure* is originally, and quite specifically, a device for a hawk to pursue and chase at his master's command, and a pounce is equally specific to falconry. A hawk will literally turn tail, present the falconer with a rear view and fly off—sometimes heart-breakingly—the wrong way.

The Battle of Maldon took place in AD 991 and the Anglo-Saxon poem about this event remains part of English literature courses to this day. Right at the start, we learn that 'Offa's kinsman' shows he is ready for battle by abandoning the falcon he has been hunting with: 'He let his beloved hawk fly from his hand towards the woods and advanced to the battle.' It's a demonstration of intent: sacrifice not only of property but of love and of joy.

There are falconry references right through Shakespeare. There's no evidence that he took part in the sport himself, but you don't have to be a great cricketer to suggest that someone has been hit for six. Hamlet says, 'I am but mad north-north-west. When the wind is southerly I can tell a hawk from a handsaw.' This is not birdwatching but falconry. The handsaw (or hansaw) is not a carpentry tool—that really would be a bit mad—but a young heron, a heronshaw: one of the hawk's prime targets.

The fact that all of life is out of joint is made clear in *Macbeth* when an unnamed Old Man says: 'A falcon towering in her pride of place was by a mousing owl hawked at and killed.' The king of birds is slain by this impossible upstart, this creature of the night, and that's so weird and so contrary as to be against nature… Worse than Viv Richards being clean bowled by a schoolboy leg-spinner.

And let's have one more. Here is Othello expressing his growing and false conviction that his wife Desdemona is unfaithful to him:

> *If I do prove her haggard,*
> *Though that her jesses were my dear heartstrings,*
> *I'd whistle her off and let her down the wind*
> *To play at fortune.*

To let your hawk go… to send your own beloved bird swooping off with the wind in its tail, a creature that you have spent hours with, sharing the difficult and complex processes of learning how to operate together, and with whom you have subsequently known pride and joy of the most vivid and lasting kind—gone, and gone forever. That's how much a good bird meant.

In the middle of the seventeenth century humans learned how to kill distant birds without the help of a falcon, and that changed everything. The shotgun was invented: a gun that could launch a tight handful of projectiles in the general direction of a bird. Smooth-bored, large-diameter weapons could be used to fire shot, and that made a radical difference to the way humans interacted with wild birds.

It wasn't an overnight process. It took time to refine the technology and make it accessible. The early shotguns weren't easy to use at all; accuracy was seriously elusive. A 'long fowler' had an overall length of over 2 metres (7 ft), not exactly handy. But with improving technology barrels grew shorter. By the beginning of the eighteenth century it was possible to take birds on the wing. That was another game-changer. Shooting birds was now a test of the shooter's skill and equipment. In other words, shooting birds was now a sport: all about fun and prestige.

By the late eighteenth century barrel lengths were shorter still and more manageable. It was possible to use a gun around a metre long, even shorter for shooting in woods. The side-by-side double-barrelled gun came into use, giving every shooter a second chance and the top

facing page:
PHEASANTS.
Never has a bird
prospered so much
because of its ability
to die.

operators a second bird. Henry Nock's patent breech was introduced in 1787, using significantly less powder and minimizing the delay between the trigger and the bang. Breech-loading guns improved. The shotgun was an affordable, accessible weapon, purpose-built for taking up arms against birds.

You no longer needed a falcon, or a falconer. You no longer needed years of training and devotion. You just bought yourself a gun.

Up to 40 million pheasants are released into the British countryside every year. Along with five million red-legged partridges. This is because pheasants are extremely good at death. Never has a bird prospered so vastly because of its ability to die. We manage most of the countryside of lowland Britain for the benefit of a bird that in natural circumstances would be no closer to us than the Black Sea.

Pheasants have two techniques to avoid predation. The first is to lie doggo and hope the danger will pass. The second is to leap into the air at the last second while making a frightful noise. This, as we have seen, is intended to provoke a startle response in the predator, who will freeze for the crucial fraction of a second it takes for the pheasant to get away.

So the most popular way of shooting pheasants is to get a line of beaters to walk through places where pheasants are hiding and to provoke them into flight, as high as possible. Once their vocabulary of escape techniques is exhausted they are at the mercy of the guns.

The agricultural landscape of lowland Britain is dotted with woods and with strips of maize. These are there to provide shelter for pheasants; as Asian birds they die from too much cold and damp. Woods that would otherwise have been grubbed up have been allowed to remain, and very often they were coppiced to provide additional cover. All this, quite incidentally, provides habitat for songbirds. Our agricultural countryside is more pleasing to the eye and to the ear as a result of shooting. It is richer in life than it would otherwise have been. The countryside has been traditionally managed—and much of it is still managed—on the principle that everything that can possibly be good for pheasants must be done.

There is a different tradition in the United States. They've always had more real wild country than Britain, and that's still the case. In America they have three kinds of landscapes, broadly speaking: town, farmland and wilderness. The towns are full of people, the farmland is an open-air food factory and the wilderness areas are pretty damn good for wildlife.

That's not how it works in most of Britain. Here we have just town and country. If we managed our countryside too intensively—with too single-minded a desire to maximize the production of food—we'd lose almost all our wildness. Shooting has played a part in making sure that the countryside in Britain has a double function: to satisfy the human need for food and the human need for wilderness.

The second principle for the management of the British countryside has been that everything that is bad for pheasants must be removed. The whole point of the countryside—after the production of food—was to produce as many fat, colourful target birds as possible.

It followed that anything with a hooked beak was wrong. An enemy. Evil. So the great tradition of Victorian gamekeeping began. The gamekeeper reared pheasants as intensively as possible, released the young, silly birds into the countryside, gave them fat rations of corn to eat and trapped or shot anything that might want to kill them. Poachers and birds of prey were twin threats to civilization and were dealt with without mercy. Ground predators like stoats and weasels were treated the same way, though foxes were often left untouched because a more exciting death awaited them. The point of the wild countryside—the great reason for its existence—was killing things.

As a result, five of Britain's species of diurnal birds of prey became extinct in this country round about the start of the twentieth century: goshawk, marsh harrier, honey buzzard, white-tailed eagle and osprey. Five more were down to fewer than 100 pairs: golden eagle, hobby, hen harrier, red kite and Montagu's harrier.

War is often good for wildlife. Humans concentrate more on killing each other than on killing birds. So the two world wars were good news for British birds of prey. There were fewer gamekeepers on the ground because many were called up, and the whole country had other priorities —and better uses for ammunition—than sport.

After both wars there was an immediate spike in the killing of birds of prey. But the game was changed in 1954, with the Protection of Birds Act. This gave legal protection to all wild birds in Britain, apart from the

sparrowhawk—particularly disliked by gamekeepers. This bird finally got protection in 1963.

It's important to mention here that gamekeepers weren't the only problems faced by birds of prey. The introduction of the disease myxomatosis hammered the rabbit population, an important target species for many birds of prey, especially buzzards. The use of organochlorine pesticides in the agricultural intensification that took place after the Second World War was also catastrophic for birds of prey. We'll consider both these matters later (see pages 249 and 252). I mention them at this point because I don't want the shooting industry to feel unfairly treated. Heaven forefend.

Shooting birds of prey is against the law and successful prosecutions are regular. In England this means that the gamekeeper gets fined and, in one recent case, sent to prison. Never the estate owner. Prosecutions are frequent enough to show that birds of prey are in some places shot, trapped and poisoned as a matter of course, though infrequent enough to indicate that the practice is dying out. The RSPB has worked successfully with the police in bringing about these prosecutions.

Birds of prey are still being killed, but the Victorian free-for-all is a thing of the past. Among game-shooting people the thinking remains: birds of prey are at best a nuisance, at worst an evil. But there are more birds of prey around in much of Britain than there have been for centuries.

This has led to a different kind of problem. It is sometimes suggested that numbers of birds of prey are 'out of control'—as if the world will cease to function unless all wild creatures are under human control. It is further suggested that we are losing songbirds because of the excessive numbers of birds of prey. Here is a new, thrilling, twenty-first-century way to make birds of prey into villains.

It ignores one of the basic principles of ecology. The predators don't control the number of prey species: it's the other way round. You don't have many lions unless there are lots of wildebeest. You don't have many sparrowhawks in your wood unless it is full of songbirds. Food supply controls every kind of population: if there is plenty of grass on the savannahs there will be plenty of wildebeest and therefore plenty of lions. If there are plenty of caterpillars in your wood there will be plenty of blue tits and therefore there might just be a single pair of sparrowhawks.

In other words, the presence of a predator indicates the health of the environment. A fabulous bit of wood will be jumping with caterpillars, so it will be full of songbirds—and so it will be able to support a predator

facing page:
THE SPARROWHAWK was particularly disliked by gamekeepers. They finally received legal protection in 1963.

or two. But note there that the predator is by definition rare. In a healthy wood there will also be far more caterpillars than blue tits and far more blue tits than sparrowhawks.

In other words, the most vulnerable kind of species in any ecosystem is the top predator.

Though the shooting of exotic introduced birds is at the heart of the shooting industry, there are some native species that are also legal targets. These include grouse. Enormous areas of upland Britain are managed to maximize the numbers of wild grouse; the aim of a day's shooting is to kill as many birds as possible. That is where the pleasures and satisfactions of the sport lie. And people are prepared to pay serious money for these pleasures: the more birds they can shoot, the more they spend. You can expect to pay £120 for each brace of grouse shot; a good day's shooting involves the killing of 150 brace. Not a cheap day out then.

It follows that everything that can be done to maximize the day's bag is seen as unequivocally good. That involves the routine and illegal killing of birds of prey. Hen harriers have been particularly hard hit. Figures from Defra (the Department for Environment, Food and Rural Affairs) show that there is room for 300 pairs of breeding hen harrier in the challenging environment of the uplands of England; in 2013 no pair successfully bred. It's hard to police in these very open and remote places, but there have been successful prosecutions. There was a famous case, in the Peak District National Park, in which birds of prey were illegally killed on land owned by the National Trust where the shooting rights had been sublet. A National Trust representative told me that, so far as he was concerned, the control of birds of prey was the way forward. This ignores the fact that (a) killing birds of prey is illegal and (b) there aren't any left to control on the uplands. Or hardly any. The great comeback of British birds of prey tends to stop dead when you get close to grouse.

This has become a cause célèbre, with the opposed sides deeply entrenched. The shooting industry attempts to paint the opposition as extremists; after all, they do take the extreme position of suggesting that shooting interests ought to operate within the law. There is an increasingly political dimension to the debate. At its heart is one big question: whose countryside is it anyway?

The violent death of birds has always had a powerful hold over the nation's emotions, and birds of prey have always been at the heart of it. At first they were the beloved beasts that did

the killing, now they are the hated enemies who spoil the fun.

In spite of this, we have better numbers of birds of prey than we have had for half a century, and that can be documented by records from the British Trust for Ornithology. It may be that there are more birds of prey, and species of birds of prey, than there have been for a century or even longer. Many species—though not the hen harrier—show a marked increase. It seems to me a pity that this is not universally seen as a good thing.

facing page:
THE HEN HARRIER is unpopular with people who run grouse moors, and its numbers have suffered accordingly.

7

Finger-lickin' good

I was woken by the sound of a cock crowing and felt oddly disappointed. I was in a hut in the middle of a wildlife reserve on the Malay Peninsular and had hoped for wilder morning sounds than a chicken. Then I remembered the previous night's journey: we had gone a long way in the dark and I had seen no lights for miles. There surely wasn't a kampong within earshot. The cock crowed again, giving permission for the day to begin, for the sun to shine, and I realized that this was no farmyard rooster. This was a red junglefowl. A wild bird. Red junglefowls are the ancestors of every chicken on the planet.

There are more chickens on earth than any other species. They have prospered because of their edibility, and because they are so easy to keep. They just get on with the job of staying alive and getting fat, eating whatever their beaks come into contact with, down around your ankles and under your feet. They seem capable of putting up with anything. There's hardly a human settlement so remote that it lacks a handful of preoccupied old chooks picking for snacks or snoozing in the shade.

Chickens provide us humans with more protein in a year than sheep, goats and cattle. They supply us with eggs on a daily basis and are always on hand when a feast day needs to be celebrated. They have been part of human life since 8000 BC: chicken remains were found among human artefacts in North China that date back to that time. All the same, they took their time spreading to every corner of the world; chickens are absent from the Old Testament, *The Odyssey* and *The Iliad*.

below:
BANKIVA FOWL.
For caption, see page 150.

How did it all begin? Why was the red junglefowl, of all the 10,000 species of bird in the world, selected to play this intimate part in our lives? If you are what you eat, then quite a lot of people are a chicken.

We like to think that the process of domestication was heroic, or at least deeply touching. It's been claimed that dogs became part of human life because wolves were captured and trained for hunting, or because wolf puppies were taken into the home for their charm. The process is more likely to have been chance-driven; human wastefulness providing free meals most days to scavenging animals, so human society acquired a sort of comet tail of followers. The wolfy ones became more biddable as time went on. They became useful to humans, at first to give warning of intruders of both the human and the non-human kind, and subsequently because their hunting senses were superior to anything we humans could offer. Dogs became the noses of the hunting human. So we began to take control over them, and bred the ones we liked best. We (mostly) put the friendliest animals together and bred still friendlier pups. Some time after

below:
SONNERAT'S
JUNGLEFOWL,
otherwise known as
the grey junglefowl.

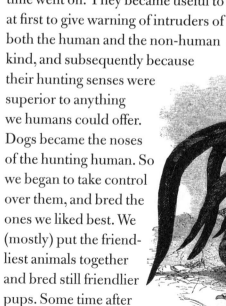

that, dogs became one of the first luxuries of human life: companions.

It's likely that the junglefowl entered human society by the same ad hoc method, at first scratching for food around settlements, tolerated because it's never a bad idea to have a decent amount of protein close to hand. So we began to encourage them: eventually to own them. Thus the partnership began—and it has profoundly affected the history of both species.

Chicken is traditionally a treat, not a staple. At the risk of sounding like the four Yorkshiremen from the Monty Python sketch, I shall recall one of my mother's tricks for making the food and the housekeeping money go a little bit further while still filling us with a sense of wellbeing. As a weekday special, she used to make chicken with all the trimmings—but without the chicken.

She made bread sauce, roast potatoes, stuffing, carrots and gravy—and served it with thin chipolata sausages. Chicken was too expensive, but sometimes in the middle of the week we had this chicken-esque meal as a sort of cheating treat: getting one over on the normal constraints of life: I can't believe it's not chicken. On Sundays we always had a proper roast dinner, usually with some cheap and uninspiring cut like shoulder of lamb, which even my mother struggled to make delicious. I suppose it was the theatre of the thing that mattered: the table with the big lump of hot protein in the middle. It said that we were a family. We were Barneses and we were doing all right. But every six weeks or so—I can remember the joy of it—we would have a whole roast chicken, and that was really a

A roast chicken made the whole family think how good it was to be alive

thing to celebrate. It was a great family treat, sanctifying our holiday and turning it into a real holy day.

Chickens were homely, sure, but they were also special. A roast chicken made the whole family think how good it was to be alive: how good it was to be all together, and how lucky we were to have the privilege of eating chicken.

Birds are hard to catch without relatively advanced technology. That's because they can fly. They are only readily available to humans once they are domesticated. The invention of agriculture was perhaps the single greatest breakthrough in the history of humankind. Humans ceased to be wanderers and became settlers. They swapped a life of uncertainty and relative idleness for one of comparative certainty and the rock-solid guarantee of back-breaking toil. Our ancestors brokered and accepted this deal, and so humans were able to spread out all over the world.

Once we became persons of fixed abode it was inevitable that chickens and other domestic fowl became part of human life: an aspect of agriculture. Birds that could live any old how, needing no great input of time and energy and feed, were a great asset, and doubly so if they carried a decent amount of meat. So other birds were brought into human lives: ducks, turkeys and—rather overlooked these days—pigeons. Pigeons are faithful to a site, very keen on breeding and expert at foraging for themselves. If you established a colony of pigeons, it looked after itself and could be harvested at will.

It was a little before the European football championships of 2008 and England had failed to qualify. So someone from *The Times* features desk rang me up and asked who I would be cheering for. No one, I said. That was the wrong answer. I was urged to pick a team—any team—and write about why I wanted them to win. It was apparently essential for the future of the newspaper that I did so. 'All right then,' I said. 'Turkey. I'm a huge fan of Turkey, all right?' Perhaps they thought I wasn't taking the whole thing with sufficient seriousness. If so, they had their revenge soon enough. 'Why Turkey?' they asked. 'Because Orhan Pamuk is a great writer, and deserved the Nobel Prize he got two years ago,' I said.

But I drew the line at posing in a Turkey football shirt; I was very grand in those days. So they asked me to go to an address in Norfolk and be photographed there. I found myself the next day on a farm that bred Norfolk black turkeys. I was pictured in the middle of a field, surrounded by hundreds of these bloody birds, all of them standing damn near waist-high, filling the air with their mad cackling and gobbling.

They were rounded up and kept in order by a swift Border collie, who used his innate sheep-herding skills to keep these monstrous birds in order. The farmer, who rightly found the whole business hilarious, picked up a specially well-sized bird and told me to hold it. So I grasped its great legs with just about the right level of firmness while the photographer snapped away. He caught me laughing all over my face at the absurdity of it all: and also at the beauty of the birds and at the beautiful way they were farmed. I haven't eaten meat since my twenties, but I recognize good animal husbandry when I see it and I had proper respect for this set-up. I had a weird kind of liking for it as well.

'And if you turn to the camera… Lift your head a bit more… Can you move the bird…?'

'Look, this bird is seriously heavy and I won't be able to hold it for very much longer. Whatever you're doing, please try to make it quick.'

He did a grand job, as it happens. The picture was published on the cover of the newspaper's second section with the caption: 'Guess who Simon Barnes is supporting at Euro '08?'

Never confuse a newspaper with a sane organization.

above:
TURKEY.
'Guess who
Simon Barnes
is supporting at
Euro '08?'

Two important facts: (1) Turkey had a great run in the competition and were beaten by Germany 3–2 in the semi-finals; (2) Turkeys have nothing to do with Turkey. The name is a classic example of bad birdwatching. Early European settlers in Central and North America confused the turkey with the guineafowl. Not so terrible a mistake, I suppose; they are both members of the order Galliformes, which includes chickens and junglefowl, pheasants, turkeys, quail, partridges and pheasants. Guineafowl are African birds, but they were imported into Europe via Turkey. So they were frequently called turkeyfowl.

Galliformes are all ground-feeders and reluctant fliers. They tend to be substantial birds, heavy for their size. Plenty of meat on them, in other words. All this made them ideal for domestication. The easiest one of all—the most tractable yet the most self-sufficient, the most well muscled without being tough, the most productive of eggs, the best of all fitter-inners—has conquered the world. In some ways the chicken defines humanity, and the bird has shaped the history of humanity, playing a major part in our transition from the nomadic to the fixed way of life. It's been there, providing eggs as a staple to mark every day and its flesh as a treat to mark the high days and holy days, for uncountable generations of humanity, and many, many more generations of chickens.

The chicken has shaped the history of humanity, playing a major part in our transition from the nomadic to the fixed way of life

We humans love a special day. We like to take a day and lift it out of the common run of days; a day that makes all the other days worth living, or, if you prefer, one that celebrates the fact that all of life is worth living. We have one such day a week—even in a secular world Sunday is a day of special meaning. And across the year we have just a few days that are even better: super-Sundays, as it were, no matter what day of the week they actually fall on. In many cultures we have chosen the turkey to mark such a day.

There is a special theatricality about bringing an entire bird to the table. An entire and enormous bird. It's not a chunk of some mammal; this is every bit, bar the guts and the overcoat, of a once-living thing, a creature that has given all of itself so that we humans might celebrate the specialness of the day and of our lives. In my childhood the entrance of the turkey on Christmas Day was one of the moments of the year. My mother would bring it in from the kitchen—no sausage nonsense on this of all days—struggling with the immense weight of it, and it would tell us of the wonders of life.

We British seldom eat turkey at any other time. It wouldn't seem right. It would spoil the specialness. To cook a turkey and present it to the family in June would seem a kind of blasphemy: an attempt to destabilize Christmas.

For the Americans, turkeys are about Thanksgiving. This is as deep and as significant a celebration for humankind as the winter solstice we celebrate at Christmas. It's about harvest. All is safely gathered in before the winter storms begin. Obviously, a successful harvest had a specially deep meaning for pioneers. A good harvest sanctified the great adventure, made the family safe for another year, made it plain that God was on your side. Of course a great bird had to be sacrificed. Of course feasting and merriment were necessary. It's a ceremony of relief, of feeling blessed. Only a turkey will do in such circumstances; anything else is too puny, insufficiently thankful, insufficiently joyous.

Thus turkeys play a vast symbolic role across America, even in modern times when most people are city dwellers and few are in touch with harvests, and even the formal measure of the seasons is less obtrusive than it used to be. Turkeys are for feasting and family: for the standout moments of the year.

Why do we call them chickens? No obvious reason, that's for sure. A chick or a chicken is a young bird still clad in down. There's a pub on Highbury Corner in London called the Hen and Chickens—the two were regarded as different things: one a grown-up, the others all babies. Somehow the term got to be used for the entire species. Perhaps we preferred a diminutive form, to show a complex mixture of affection and contempt.

Females are hens, we're all agreed on that, but what do you call a male? That's traditionally a cock, but a cock is also a term for the male member, and rather a poetic one, a bird-cock (a male chicken) being a colourful, exuberant and perhaps slightly preposterous expression of masculinity. The figurative cock out-competed the literal one, and so in America you don't have cocks at all. You have roosters. (The same bowdlerization exists in the way Americans reckon weight: they work in pounds rather than stones, because 'stones' is an ancient—and now obsolete—term for testicles.) All chickens roost, so they're all roosters, but language doesn't develop logically or literally.

In the meantime, the term 'cock' has become an uneasy one in British English as well. As a result, 'rooster' is sometimes borrowed from

America, but more frequently we use 'cockerel'. Technically, a cockerel is a male bird under the age of one, but now any hoary old farmyard cock is likely to be called a cockerel. Which seems to me a bit of a rooster-up.

How many chickens are there in the world? One estimate—by no means the highest—gives you 18 billion, or three for every person. Chickens are reared in such prodigious quantities because of a revolution in agriculture. It's usually termed factory farming. It's about the divorce of meat production from the natural environment.

About three-quarters of all the chicken meat consumed in the world comes from factory farms, and about two-thirds of the eggs. A chicken can live for five to ten years; industrial chickens live for six weeks from egg to table.

The process operates by means of a profound mental shift. You stop thinking of chickens as animals and start treating them like plants. You keep them in a very small, confined space and give them the liquids and nutrients required to bring them to full ripeness. Then you harvest them. Any notion of a chicken's natural mode of existence is set aside. It's a bit like hydroponics: an ingenious and effective way of growing plants without soil, in buildings designed for the purpose. There are online guides to growing hydroponic marijuana.

The critical issue in factory-farmed chickens is space. In some operations you will find up to twenty birds to a square metre. Any movement at all is discouraged. Frequently the ends of the beaks will be clipped off to discourage fights. The birds sit in their own ammoniac waste for six

A chicken can live for five to ten years; industrial chickens live for six weeks from egg to table

weeks before they're slaughtered. There is a pre-harvest mortality rate of 3–5 per cent.

A great deal of this factory-farmed chicken meat goes to the fast-food industry. Some more facts: one-third of Americans eat fast food every day; two-thirds of Americans are overweight; one fifth are obese. In other words, it's now normal to eat more food than the body can usefully accommodate. The fast-food industry worldwide spends 20 billion US dollars a year on advertising.

A chicken meal has gone from a treat to a daily staple. It has gone from a luxury to something seen as a basic human right. It has gone from a healthy option to an unhealthy one. Chickens were once a great life-enhancer. Now they're part of a global health crisis.

Chickens tell the history of humankind, from the glorious optimistic beginnings of civilization to a decadent and self-destructive triumph of technology and development. Chickens helped humans to put down roots, to have permanent dwellings, to settle down, to build, to think, to develop, to breed and spread and prosper. Now they are part of the flight from nature, the divorce of humanity from food production, the turning of animals into objects. Chickens are no longer birds, they're McNuggets. Chicken is now a term for stuff we put in our mouths, its form and shape determined by human minds. We've found the bucket; we've lost the bird.

8

Symbols at your door

M y dear Aunt Barbara used to tell a story about a vicar —I have long forgotten his name and parish—who took evensong after a protracted and agreeable lunch. With his belly full of claret and port and other vicarly delights, he approached the lectern but, alas, misjudged his descent from the steps of the altar. He made a dramatic lurch in the general direction of the congregation but saved himself by clasping the outspread wings of the lectern, for this traditional piece of ecclesiastical furniture was, of course, in the form of an eagle, its wings supporting the Book. He muttered, in tones audible to the front row, 'If it hadn't been for this bloody duck, I'd be on the floor.'

The symbolic role played by birds in human culture is so vast and various and universal that it would fill several volumes; indeed, it has done so on many occasions. My plan here is not to be encyclopaedic but to seek out a couple of examples that will give a flavour of the power that birds have over our minds, affecting our language, the way we think and the way we see the world. So I'm not offering a lightning survey of the meaning of every species in every culture. Here instead is a pair of birds so heavy with the weight of symbolism that it's a wonder they can still take off.

Let's consider the eagle and the dove. Which is pretty much like saying war and peace, death and life, heaven and hell, hope and despair,

previous page:
BALD EAGLE.
For caption, see page 165.

horror and joy. But never in a wholly straightforward way. These are highly complex ideas and the symbolism associated with them is never simple. Sometimes these symbols seem utterly clear and uncompromising, but at other times they are bewilderingly ambiguous, seeming to combine several meanings at the same time, some of them disturbingly contradictory. That's when you start to wonder if the symbol is more potent than the meaning, overpowering the ideas that lie behind it by means of its glorious vividness and profound variety of meaning.

Eagles have been used to represent the Holy Spirit and Nazi Germany, along with many other things. The white dove is a sign of peace. White doves are artificially bred from ancestral rock pigeon stock... the same species as the feral pigeon, a bird cursed daily by a billion city dwellers and called a flying rat. No symbol is without potent shape-shifting ambiguities. No one—neither God nor Hitler—has a symbol on his own terms.

The eagle was the symbol of Zeus. So eagles are about power, then. And majesty. Power over gravity, power over the things that live on the earth beneath. An eagle is capable of seeing things from a great distance above, and then swooping down and carrying them off. Pliny the Elder was capable of accurate observation, and equally capable of moving seamlessly into—in this case literally—the most high-flown and speculative of ideas. Thus he notes that the eagle is the only bird that has never been struck by a thunderbolt. This is because the eagle is the bird of Zeus (though as a Roman he knew the deity as Jupiter) and therefore

facing page:
EAGLES
command the attention.
They force you to
look up.

Zeus, being also the god of thunder, makes sure his thunderbolts fall a decent distance from every eagle in the world.

Ganymede is described in *The Iliad* as the most beautiful of mortals, and Zeus, who always kept an eye on the human world for anything he might fancy, took a great shine to him. So he assumed the form of an eagle and carried the lovely boy up to Mount Olympus to serve as his cupbearer—a kind of celestial barman. Ganymede was given the gift of immortality and eternal youth. The liaison between Ganymede and Zeus represented a socially acceptable erotic relationship between a male youth and an older man.

So you can, if you wish, see the eagle as a symbol of homosexuality instead of, or as well as, a symbol of Nazi power. Or the tumbling vicar's bloody duck.

Eagles command the attention. They force you to look up; you can't help it. The least nature-conscious person in the world looks up at an eagle. The nature-blind see an eagle and raise their eyes, suddenly struck sighted, in a twist on St Paul's Damascene experience (something that you can read about in the big book that's perched on the eagle's wings).

I have been on Mull, an island off the west coast of Scotland, and there I watched the eagle-watchers. Few of them have binoculars. Here, at a designated eagle lookout, there are likely to be good people from the conservation organizations with binoculars to lend, so this may be the first time their visitors have ever looked at a bird through a pair of bins or

ever seen a bird close up. They are here and looking hard because eagles command the human imagination.

The white-tailed eagle became extinct in Britain (see page 173) but was reintroduced in the 1990s and has since prospered. Mull has prospered with them; the eagles are one of the island's great tourist draws.

Even in places full of eagles, the eagles stand out from the rest of the wild world. In Zambia the call of the African fish eagle announces the evening news on television, and naturally an eagle flies on the country's flag and on its coat of arms. Eagles across the world, in every culture, are irresistible to humans. And it's a process that begins with their appearance. All the associated meanings—power and savagery and majesty—come second. It begins with a non-verbal, unthinking, gut-deep response to a flying monster cruising overhead.

And I think that, at bottom, this response is one of longing. Everyone who looks up wants to be that eagle; wants to be like Zeus and assume the form of an eagle. We envy all forms of flight, but the eagle rubs it in harder than anything else. Because eagles are so big—the wingspan of a white-tailed eagle can be 2.2 metres (7 ft 3 in)—they make the fantasy of flight look attainable. Almost…

In Zambia the call of the African fish eagle announces the evening news on television

What is an eagle? That's not a straightforward question, particularly when we are talking about symbolic eagles. For a start there is a clash between vernacular eagles and ornithological eagles. A bateleur eagle is a supreme flier and as inspiring a bird as anyone would wish to see—but it's not a 'true' eagle in strict taxonomic terms. It is related to the group called 'snake eagles', so called because snakes are an important part of their diet. The great martial eagle that flies over the same savannahs of Africa is a different matter, as true an eagle as you could wish. Both are head-turners, both fabulous fliers, both memorable sights. One has completely feathered legs and is an eagle; the other doesn't and is a snake eagle. A true eagle wears long trousers, while a snake eagle goes barelegged. That matters to a scientist; it also matters to a birdwatcher, at least to an extent. But it doesn't matter a jot to most people. And it is a thing of total irrelevance so far as myth-making is concerned. Classification doesn't affect the symbolic meaning of a bird. Nor do trousers.

In Scotland many people thrill to the sight of a golden eagle: and quite often it's actually a buzzard, which is a kind of hawk. Does that diminish the experience for them, I wonder? Is it like looking at a reproduction of the *Mona Lisa* instead of the real thing in the Louvre? Is one experience more meaningful than another? Who's to say?

In St Matthew's gospel, Jesus says in the Authorized Version: 'For wheresoever the carcase is, there will be eagles gathered.' This is elsewhere translated as 'vultures'. There are certainly vultures to be found in Israel, though the most common carrion feeders are black kites. But who knows the strength of the local populations and the species mix in the first century AD? There weren't many ornithological surveys being done back then, that's for sure. So any nice big fierce bird can be called an eagle in some circumstances.

Humanized and symbolic eagles are different from the eagles of field guides. As the human mind hunts for ancient and archetypal meaning in what is all around, so the real world becomes soft-edged and protean. It's not about feathery trousers. If you want it to be an eagle, it's an eagle.

Eagles glide gloriously in and out of all mythologies. Vishnu's chosen mount is Garuda, who has a man's body and the wings of an eagle; he is also renowned as a serpent-swallower. Garuda, then, is a righteous creature, fierce and uncompromising in the face of evil. The Aztec sun god Huitzilopochtli takes the form of an eagle. So does the Nordic storm god Thiassi. You can find eagles, along with other kinds of large and imposing birds of prey, on every continent except Antarctica. Eagles, condors, vultures, kites, buzzards—they are inspiring birds and humans have found inspiration in them wherever civilization has been established. Eagles own the sky: the heavens, if you prefer. They seem to own the entire world. Perhaps they are the most powerful and most nearly universal creature in the lexicon of human symbolism. They are part of our history, but they go much deeper than that. They are part of our mythology as well. They are part of what made us, and part of what sustains us.

The Aztec sun god Huitzilopochtli takes the form of an eagle. So does the Nordic storm god Thiassi

Perhaps eagles were the first logo. The Roman Empire was very keen on brand recognition and they adopted the eagle—Jupiter's bird, remember—as the symbol for their military might. The bird was generally shown above the letters SPQR: *Senatus Populusque Romanus*— the senate and people of Rome. Each legion carried its own standard, which took the form of an eagle, and its loss was a terrible disgrace.

Naturally enough the eagle was adopted into the medieval notions of heraldry. A shield bore the arms of the person who carried it, and it was an expression of his power and his pedigree. Britain was keener on lions than eagles—the arms of both Scotland and England carry lions—but eagles still thrive on the most famous escutcheons of the continent. In heraldry the eagle is normally presented spread out, as if flattened on to a table for vivisection, so as to show off its wings and its talons—both fierce and helpless at the same time. The heraldic term for this position is 'displayed'. The eagle displayed can still be found on national badges, crests, flags and coats of arms from Albania to Zambia.

Charlemagne pre-dated heraldry, but he was given the retrospective coat of arms of an eagle as befits the first Holy Roman Emperor. It was carried by all who followed him. In 1211 the eagle acquired a second head when it became the symbol of the Byzantine Empire. The eagle that looked both ways showed the world that here was an empire that looked to both Europe and Asia—to both Rome and Constantinople—and so out-flew the bird of Charlemagne.

The single-headed eagle appeared again when Napoleon I adopted it as his personal symbol. He filched the idea from the Romans, with whom he identified so strongly. He also borrowed the idea of the portable eagle as a regimental standard.

The United States also took on the eagle as a national symbol, though this was no generic eagle. The bald eagle is a proper species (also a true eagle, so far as that matters) and pretty distinctive as well. True, the African fish eagle looks very similar, but best not mention that

facing page:
THE BALD, OR
WHITE-HEADED,
EAGLE.
Between 1917 and 1952
128,000 were shot in
Alaska alone.

while you're in America. It's certainly a stunning bird in reality, and has become part of American iconography as much as or more so than the image of Uncle Sam. You can find the bald eagle on a 25c coin (call it a quarter if you don't want to be laughed at) and on banknotes.

Prussia adopted the eagle as its own symbol, and that was adopted by all of Germany under the Weimar Republic of 1919 to 1933. The eagle was then claimed and modified by Adolf Hitler and it became a secondary symbol of the Nazis. This particularly ferocious eagle was often portrayed carrying a swastika in its talons. You can say what you like about Hitler, but he was a master of branding; those that follow—McDonald's, the Soviet Empire, Coca-Cola—can only admire and envy. The eagle was then modified to become a symbol of post-war Germany. It became less skeletal and less vicious, an eagle capable of compromise. This less belligerent bird is sometimes referred to as *die fette Henne*, the fat chicken.

Power. The might of a nation and its ruler. That's been part of the eagle's symbolic meaning across the centuries, and it's not terribly edifying to twenty-first-century eyes. The eagle has been used as a summons to blind patriotism and unambiguous hero worship, on occasions raising its bearer to something close to sainthood or even divinity. It's as if the symbol of the eagle gives a spiritual power: a divine thumbs-up to the person who dares to adopt it. To recruit an eagle seems in some circumstances to be more or less the same as having God on your side.

But eagles are about God as well, just as much as they are about megalomaniacs.

When you concentrate on the flight rather than the ferocity of the eagle you find a symbol not of war and conquest, but of religious aspiration: a vast bird that flies as if the air were a sofa, climbs to vast heights without so much as flapping a wing and cruises across the land as if gravity were a matter of no account. Clearly such a creature is an intermediary between earth and heaven: a sign of human aspiration to higher things, and also of God's infinite generosity in reaching down to the flawed creatures who inhabit the surface of the earth. By the time we reach the New Testament the eagle has become a symbol of the Holy Spirit.

In Exodus God says to Moses after the affair of the Red Sea: 'Ye have seen what I did to the Egyptians, and bear you up on eagle's wings that I might bring you unto myself.' (Hard luck on the Egyptians, perhaps, though it's interesting to note that there was an eagle on the seal of the Ptolemaic pharaohs.)

Isaiah rendered into Jacobean English brings us some of the most uplifting stuff ever written, so naturally we can find an eagle or two here: 'But they that wait upon the Lord shall renew their strength: they shall mount up with wings as eagles; they shall run, and not be weary; and they shall walk, and not faint.' The priest and composer Michael Joncas put together a number of these references in the devotional song 'On Eagle's Wings', which is sung in churches of many denominations, demonstrating that even in a century in which few people have seen a real eagle, the eagle's imagery remains vivid and important.

Even in a century in which few people have seen a real eagle, the eagle's imagery remains vivid and important

Eagles operate as a fierce aspect of spirituality. For some spiritual purposes, all eagles are snake eagles: emblems of nobility and aspiration that regularly cleanse the world of the evil and poisonous things that creep across its surface. The eagle is the symbol for St John the Evangelist, which is one of the reasons for the eagle lecterns. There is a painting in the National Gallery in London by Domenico Zampieri that shows St John receiving the divine inspiration to write his gospel. His eyes are turned heavenwards while his feet rest on an eagle. It seems that the evangelist's aspirations have already risen higher than eagles can carry him.

Spirituality, then, is not a fuzzy New Age sort of thing. It is not entirely lovey-dovey, for all that we will be coming to doves very shortly. The eagle symbolism demonstrates something hard-edged and direct—imposing, and not a little frightening. There is an eagle flying around the throne of God in the Book of Revelation, the last book of the Bible and the most alarming.

Eagles represent religion at its least compromising. They carry a sense of magnificence and aspiration, but also an aura of hardness, of the cruel-but-necessary kind. Like all symbols they are packed with ambiguities of every kind.

The Bible has its jaunty, even its saucy, moments. Eagles aren't always connected with high and lonely duties of the spiritual quest. Sometimes they can be persuaded to play a more earthy role. There is a chunk in Proverbs, one of those bits of the Bible half-known to us all, though few of us can remember where it actually came from without looking it up. I did so myself, of course, and was delighted to be reunited with this verse. Savour the rhythm of it:

> There be three things which are too wonderful for me, yea, four which I know not: the way of an eagle in the air; the way of a serpent upon a rock; the way of a ship in the midst of the sea; and the way of a man with a maid.

The white dove is a symbol of peace, love and fertility, the three things being naturally connected. It also stands for the Holy Spirit among its many other religious duties. So what is the difference between a dove and a pigeon? None whatsoever. We use the term in an apparently random way, but ultimately, the species we like we call dove and those we like less we call pigeon. Thus the bird of our cities is variously known as the feral pigeon and the common pigeon; both come from the same wild species, which we call rock dove.

The domesticated pigeon is, as said before, bred from the same species. The homing pigeons, fantails, pouters and tumblers that pigeon-fanciers love are blood brothers and sisters to the birds we curse in city streets. And so the white dove that we love is the same bird that scavenges for chips outside McDonald's at Liverpool Street Station in London.

In James Joyce's great novel *Ulysses*, there is a reference to Léo Taxil's blasphemous book *La vie de Jésus*, in which Joseph talks to Mary about her pregnancy. 'Qui vous a mis dans cette fichue position?'

'C'est le pigeon, Joseph.'*

It's funny because Mary uses the word 'pigeon' rather than 'colombe'. Or dove. That's the joke that brings banality to this otherwise sacred matter. A dove can be profound, but to turn it into a pigeon is a belly laugh. Thus our ambivalence about birds, both real and symbolic, is neatly skewered.

Doves were religious birds long before Christianity. How could they not be? Rock doves are enthusiastic cave-nesters, and so were associated with the very beginnings of human society. Their tranquil cooing told us that all was well. They are readily tamed and, for that matter, killed—sacrificed, if you prefer—so they supply food as well as companionship. They are generally seen as half of a pair; the phrase lovey-dovey isn't just there for the rhyme. They eat seeds rather than birds or mammals. They go away and they come back. They are birds associated with good

* 'Who put you in this damn position?'
  'It was the pigeon, Joseph.'

qualities and gentle times.

Thus symbolic doves were associated with Aphrodite, Ishtar and Astarte. In *The Epic of Gilgamesh*, the Sumerian poem that is reckoned to be the world's first great work of literature, a dove is released to find out if the great deluge has ended.

In a curious echo, in Judaeo-Christian culture we first meet doves in the eighth book of Genesis, when Noah sends a dove from the ark to look for the land, to see if the great flood has at last relented. The dove, faithful in the manner of doves, comes back with nothing to show for the journey. But seven days later, Noah tries again, and this time the dove flies back bearing in its beak a fragment from an olive tree, and Noah knew 'that the waters were abated from the earth'. So he went on to kick-start humankind's second attempt at creating a fair, just, peaceful and gentle civilization.

The dove's burden has been translated in the King James Bible as 'olive branch', though a branch would be a hefty load for a mere dove. Later translations prefer 'leaf'. But, of course, it's the olive branch that has gone deeply into our language, and across newspapers today the leaders of nations and of football clubs routinely offer each other olive branches with varying degrees of sincerity.

Thus the dove carrying a chunk of olive tree—normally represented pictorially as a leafy twig, a handy compromise—has become a global symbol of peace.

We live in the country of the hawk but we do so in the perpetual hope that we will soon be moving on into the land of the dove

The word 'dove' is now a part of routine political slang or shorthand. Anyone who is set on a policy of appeasement, non-confrontation and compromise is a dove, while those espousing the opposite tendency are hawks. Not eagles, interestingly, though the eagle is, as we have seen, unquestionably the world's top choice as the bird of war. I suspect that's because to refer to the hawks as eagles would give them an unfair advantage, implying nobility, loftiness of aim and royal connections to boot. Hawk is a more all-purpose word, and so warlike politicians are hawkish rather than eaglish.

Either way, birds of prey and doves capture a classic symbolic antithesis, each one acquiring from the other greater vividness and deeper meaning. And, as always, these symbolic meanings don't reveal a great deal about the birds themselves and the lives they lead, but they say a very great deal about humankind. We are a peace-loving species that is forever going to war. We live for peace, but we are never convinced we can find it without going to war first. We live in the country of the hawk—or the eagle—but we do so in the perpetual hope that we will soon be moving on into the land of the dove. As soon as everyone else sees the world exactly as we do, the dove will dominate. Until then, we must live with eagles and hawks.

Doves also represent the Holy Spirit. In Luke's gospel, Jesus is baptized by St John the Baptist: 'and the Holy Ghost descended in a bodily shape like a dove upon him, and a voice came from heaven, which said, "Thou art my beloved son; in thee I am well pleased".' There are paintings that show this event quite faithfully, a half-stripped and soaking Jesus with a white dove above.

Early Christians adopted the dove as a symbol, again almost as a kind of logo, often with the word for peace attached. Doves turn up in the Psalms, perhaps most memorably in Psalm 55, which contains the lines:

Oh, that I had wings like a dove! I would fly away and be at rest. Indeed, I would wander far off, and remain in the wilderness. Selah. I would hasten my escape from the windy storm and tempest.

Doves and eagles are part of our symbolic landscape. They help us to make some kind of sense of the world that we live in. And even though we now live in times when relatively few people have seen an eagle and in which doves—feral pigeons, if you prefer to call them by one of their other names—are seen as a nuisance and a health hazard and great sums of money are spent on their extermination, we still turn to the images our ancestors chose to understand ourselves and the way we live. There may be vastly fewer real birds now living out there in the real world, but in the packed and complex world of human symbols, birds are thriving as well as they ever have. The doves and eagles of the imagination are part of everybody's lives.

The British have always been enthusiastic persecutors of eagles, and remain so to this day

Real eagles and real doves are another matter. Eagles may thrive in the human imagination, but for years the real ones have been regarded as vermin, to be destroyed on sight by whatever means possible. This is as curious a contradiction as the human mind can come up with: but then holding two opposed beliefs at the same time might be regarded as the bedrock of human civilization.

No single species is revered as much as the bald eagle is in America. It is, as we have noted, a truly special bird: recognition of this fact is part of being an American. That doesn't let real bald eagles off anything that humanity can throw at them. We are prepared to admire eagles until we see them as rivals for the same food resource, and after that there's no holding back. Bald eagles were systematically shot during the first half of the twentieth century, with a bounty of up to $25 paid to the marksman.

Records show that between 1917 and 1952 128,000 bald eagles were shot in Alaska alone, mainly because they took the salmon that humans wanted for themselves. It's reckoned that the number is actually closer to 150,000, with many bounties going unclaimed. A report from 1970 found that a single ranch in Wyoming shot 770 bald eagles, all at 25 bucks a pop.

Broadly speaking, there are two ways in which humanity responds to such a crisis: by reacting when it's (a) too late or (b) nearly too late. In this case, they took the second option. By this time there were problems other than shooting affecting bald eagles; we will examine the effect of pesticides on bird populations later on (see page 249).

In 1973—modern times—the bald eagle at last received legal protection under the Endangered Species Act, and since then its numbers have risen again. By 2007 they were taken off the endangered list in the United States, but they still receive protection. There are now around 100,000 bald eagles found across North America, with around half of them in Alaska and British Columbia.

The British have always been enthusiastic persecutors of eagles, and remain so to this day. The white-tailed eagle was persecuted to extinction in this country: the last one was shot in 1918. But they hung on elsewhere in Europe, and towards the end of the last century a reintroduction programme was started on the Isle of Rum. They have gradually spread out from there, and, as noted, they play an important role in the tourist business of the Isle of Mull (see page 160).

However, their opportunities of spreading out from there are

compromised by the fact that this brings them into the grouse-shooting estates, where unabashed illegal persecution continues to this day.

The same is true of the golden eagle, which is Scotland's national bird and an important tourist draw, quite apart from its own claims to a place in this world. The grouse-shooting industry, as we have seen, is about big bags and big money, and it is backed by a number of very influential people. Eagles may be treasured by most people, but when they get in the way of a wealthy minority, they get shot. Symbolic eagles are cherished: political eagles haven't got a prayer.

There is a report of an incident in South Ontario in 1866, when a flock of passenger pigeons flew over. The flock was 1 mile (1.5 km) wide, and an estimated 300 miles (500 km) long. It took 14 hours to pass and it contained—so far as anyone could reckon—3.5 billion birds.

Handsome things, passenger pigeons, the males a fetching mixture of slate-blue and copper. This was a flock of doves that contained uncountable millions of individuals. It's reckoned that a quarter of all the birds in North America were passenger pigeons. They were probably the most abundant species of bird in the world: perhaps the most abundant bird species there has ever been.

Their strategy was based on vast numbers and rapid movement, shifting from one superabundant food source to another. They travelled in such crowds that predators could scarcely dent their numbers. They were a classic example of evolutionary success: a dazzling achievement, nothing less.

Not half a century after the great flock flew over Ontario the passenger pigeon was extinct: one of the most notorious extinctions of historic times. That shows how much reverence we have for real doves. Many of them were shot to provide cheap food for American slaves.

The strategy that had worked so well across the centuries became a liability when humans took an interest. Their vast numbers made them easily harvestable. You didn't have to be much of a marksman, after all. To add to the problems, the birds were also 'harvested' at their vast nesting sites. Thus the birds were caught in a pincer movement and were slaughtered with immense energy until there was none left in the world. The last passenger pigeon died in Cincinnati Zoo in 1914. Her name is famous in certain circles. Martha: the dove that died.

More than a century on we're still shooting doves out of the sky as if there was an inexhaustible supply. The turtle dove is as pretty a bird as anyone could wish for, and not to be derided as a mere pigeon. Its

facing page:
THE PASSENGER
PIGEON.
Its eradication
is one of the most
notorious extinctions
of modern times.

voice is lovely: a deep purring from the treetops that seems to speak of peace and eternal contentment. I have always imagined it as the voice of the Edwardian idyll: the bird that sings while the church clock stands at ten to three, there's honey still for tea… and no harm can ever come to a contented world.

They are birds that have a famous line in the Bible, even if it's one that has often perplexed: 'For lo, the winter is past, the rain is over and gone. The flowers appear on the earth, the singing of the birds is come and the voice of the turtle is heard in our land.' The turtle is the turtle dove, scientific name *Streptopelia turtur*, the last word being onomatopoeic. *Turrr! Turrr!* The dove's in his tree and all's right with the world. The lines are from the Song of Solomon, and it was some years before I realized that this wasn't some mad fancy of singing reptiles.

We have lost 93 per cent of our turtle doves since 1970. In that time they have gone from commonplace birds, from ambient birds, from birds of the background, to very special birds indeed. When you hear one, you boast about it, and if you are involved in practical conservation, you report it.

It's an astonishing decline, and a great deal of this is the result of shooting, though there are also problems with habitat destruction and loss of the birds' favoured seed-bearing food plants. The birds are shot on their migration, especially in Malta. They funnel over Malta to make the sea crossing less perilous, and often stop over to refuel. So the gallant sportsmen of the island shoot 'em down. Turtle doves are also shot in Greece, Italy, Spain, Cyprus and France, but the Maltese are probably the best at it.

Peace, it seems, is as elusive in the real world as safety is for a real dove. A symbolic dove is peace; a living dove is a target for a man with a gun

It's a hot political issue in Malta: there are annual protests, there is international pressure, there is pressure from the European Union (for Malta is a member even though it prefers not to adhere too strictly to European laws on conservation). It's been reckoned that turtle doves will be extinct by 2021.

All this reinforces the point that there is a difference between a symbolic bird and a real one. We are happy to let symbolic birds thrive in the safe world of the human imagination, but birds that live in the real world are faced by innumerable perils. And not the least of these perils is human stupidity. Bob Dylan asked: 'How many seas must a white dove sail, before she sleeps in the sand?' Peace, it seems, is as elusive in the real world as safety is for a real dove. A symbolic dove is peace; a living dove is a target for a man with a gun. The Holy Spirit comes down in the form of a dove on a stained-glass window, while real doves are being wiped out as a bit of fun.

You can say what you like about the human race, but we're pretty damn good when it comes to contradiction.

9

*He stuck
a feather in
his hat*

A feather is strictly for the birds. Or strictly for the theropod dinosaurs, if you prefer. This last group, the feathered dinosaurs, are the ancestors of all birds: birds are the lineage of dinosaurs that survived the great extinction.

Either way, a feather is a miracle. It makes flight possible, which is miracle enough to be going on with, but it is more or less agreed that feathers didn't actually evolve for flight. Flight by means of feathers was nothing more than a glorious accident: one of those absent-minded miracles of evolution. For it's generally accepted that feathers evolved as insulation. Flight was just a handy add-on, but one that quite literally gave the dinosaurs / birds a new direction. That is to say, up.

In other words, something that evolved for one thing turned out to be frightfully good at something completely different. This happened by pure accident. There was no idea of purpose or perfection involved. It was just dumb luck. And that, of course, is one of evolution's running themes. The lungs with which you breathe as you read these words evolved from a device that is used to keep fish stable in the water without moving. It's called the swim bladder. We are able to live on land because fish developed a device that allows a goldfish to stand still in the middle of his goldfish bowl. So no, the dual use of feathers is not particularly unusual. It is, however, uncommonly dramatic.

previous page:
Superb bird of paradise.
For caption, see page 186.

Broadly speaking, there are two kinds of feathers in modern birds. The vaned feather or contour feather is what we normally think of when we hear the word: the feathers on the outside of birds. Each one has a central spine called the rachis, which carries the strands that are called barbs. Each barb carries smaller barbs called barbules, which in turn carry barbicels. Feather maintenance is a constant preoccupation of birds; these various strands must hold together like a zip if they are to be effective for both waterproofing and flight. That's why birds at rest are so often seen fussing about with their feathers. We have given the word 'preen' a pejorative meaning: a person who preens is one who thinks far too much of himself. But for a bird, preening is second only to breathing when it comes to life support.

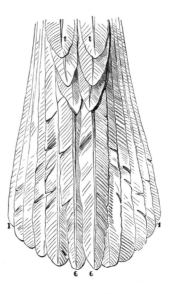

left:
TAIL FEATHERS.
The feathers which cover the roots of the tail feathers are called coverts (t).

below:
WING FEATHERS of a falcon (a) and a sparrowhawk (b).

The other kind of feather is down. Down feathers don't have barbicels, so they don't zip together. Instead, they fluff. As a result of this, they are outstandingly efficient at trapping air, and that makes them superb insulators. The snag is that they stop functioning when they get wet. So the down feathers lie underneath the vaned feathers,

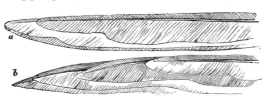

and the vaned feathers are kept waterproof with oil secreted from the bird's preen gland. That's why water rolls off a duck's back like water off a duck's back.

There's a third kind of feather called a filoplume. It is, as the name suggests, simply a strand, more or less hair-like, sometimes with a few barbs at the tip. These often seem to be pretty well useless save as decoration; some birds of paradise include these in their lavish plumage and use them when they display to a female. But it's been suggested that filoplumes play a useful part in the bird's awareness of its body position.

For a bird, looking great is an essential part of their strategy for living on this planet

Feathers are useful things. They are also highly decorative. They are used for camouflage, and in two different ways. The first is cryptic camouflage, in which a bird takes on the colours of the environment and appears to melt into them: a nightjar sitting longways on a branch is almost impossible to make out.

Feathers are also, rather more mysteriously, used in disruptive camouflage, with patterns of colouration that make the observer lose track of the bird's outline. The great spotted woodpecker is a fine example here. When one of these comes to eat the peanuts on a bird feeder it stands out like a bar of soap in a coalscuttle, but when it returns to the branches you can no longer make out its shape. The bird effectively disappears, and you can't separate it from the patterns of light and shade in the canopy. Thus they put into action the words of the zebra and the giraffe in Rudyard Kipling's *Just So Stories*. The pair, revelling in their newly minted stripes and blotches, tell the leopard: 'One—two—three! And where's your breakfast?' And vanish.

But as we have seen, feathers are also used for display: to attract mates. A male bird in prime condition will tend to look great, and many species take on flashy colours—especially male birds competing for mates—to emphasize that fact. For a bird, looking great is an essential part of their strategy for living on this planet.

Even in cases where the sexes are similar, it's hard not to speculate about a mutual sizing-up process. There is also the business of species recognition going on here, as mentioned earlier. Birds like to make it quite clear that they are members of the species they belong to, so there is no chance of getting confused. That way they will be sure of mating with a bird of the same species. Colourful feathers are part of the continuity of life in many species.

Feathers, then, are desirable objects. It follows that humans have always desired them. We have done so throughout history and we continue to do so today. There is a famous image in the Egyptian Book of the Dead in which the jackal-headed god Anubis weighs a human soul against an ostrich feather: heavier souls were devoured, while souls of sufficient lightness were permitted to enter the realm of the dead.

The weight of feather, it might be said, is what tipped the scale of human history and changed human attitudes to the wild world. A great deal of our modern awareness of the fragility of the wild world and the limited nature of its recourses was made clear to us by our crazy need for feathers. We'll get to that shortly.

Feathers have been used by humans throughout history for clothes, for warmth and for decoration. They have also been used for fletching arrows. The feathers on an arrow give it stability in flight, which means that, all things being equal, an arrow will go where the archer intended. Fletcher is a traditional English surname derived from those who practise this skill of making accurate arrows.

Most of us pick up a pen many times a day without realizing that the word 'pen' means feather, from the Latin *penna*. A feather with a suitably stiff rachis makes an admirable pen, the flight feather (one of the primary feathers) from the wing of a goose being preferred. A swan's feather is sometimes used for more ornate passages. I write a somewhat speculative italic hand myself, and once made myself a quill pen with the aid of a sharp knife. (That's why we call a pocket knife a penknife; you can use it to make a pen and also to renew its edge, for they are soft things and wear down fairly quickly.) The stroke—the feel—given by a proper quill is incomparably better than any metal nib; it bends sympathetically with the pressure of your fingers and is infinitely more expressive. The drawbacks are considerable, though: it blots easily; even with a reservoir it doesn't hold a great deal of ink; and you need to renew the point (for italic script you make a chisel point, which delivers both thick and thin strokes, and suits the quill best) and the split along which the ink flows.

A feather with a suitably stiff rachis makes an admirable pen, the flight feather from the wing of a goose being preferred

When we think of feathers for personal adornment, we tend to think of the Plains Indians. Most of us have seen feather-wearing Native Americans in a thousand westerns, as well as historic images in more authentic circumstances. The war bonnet—a headdress with many feathers—has long been a fascination for observers. It stands for alien civilization, savagery, nobility, dignity, wildness, threat and fear. It is an extension of a practice among these people, in which those who performed a great deed were rewarded with an eagle's feather; again, usually the primary feather from the wing, the biggest feathers on a bird's body. Thus those who had performed many great deeds were able to demonstrate the fact by wearing a many-feathered war bonnet on ceremonial occasions. The idea that people actually went to war in those things is one of the many fantasies that have been imposed on Native American culture; it's hardly the most practical headgear for combat.

The principle that people can read something of your history and status from such obvious clues is not restricted to Plains Indians. When I was a member of the Wolf Cubs, I managed to acquire five proficiency badges, which carried the same sort of heroic meaning, at least to me. We have already touched on heraldry in this book: a coat of arms shows your own achievements and those of your ancestors. The badge of the Prince of Wales comprises three ostrich feathers inside a coronet, a mark of honour for holders of that title that dates back to the Black Prince. The wearing of an Old Etonian tie is another classic example of the same need for self-advertisement. The Plains Indians preferred the feathers of eagles. It's all in the way that tradition takes you.

Many other people and many other places have revelled in the power and significance of the feather. In the Museum of Ethnology in Vienna you can admire Montezuma's headdress—so called, though the exact provenance is unknown. Certainly it is an Aztec artefact, acquired at the

time of the Spanish conquest of Central Mexico, and certainly it is gorgeous. The star part is played by the feathers of the quetzal, a bird even more gorgeous when actually living. It is a forest species, and the males are splendid, their long green tail feathers being particularly memorable. Quetzalcoatl is the name of a local deity. He is the feathered serpent, a snake that wears the feathers of a quetzal, so he is sinister as well as beautiful.

Montezuma's headdress has other feathers besides those of the quetzal and they are all mounted on gold that is studded with precious stones. It is interesting that the gold plays a supporting role, literally, to the lovely feathers, and that the precious stones likewise play a secondary role. The feathers are the thing: more beautiful than any dead stone, infinitely more meaningful, and a great deal more powerful, certainly in the associations they have for us humans. The feathers are about life: though a good number of lovely birds had to be killed to make the point. And ultimately the point is, of course, power.

Sir David Attenborough has always had a special fascination with birds of paradise. Two of his earliest programmes for the BBC were *Quest in Paradise* and *People in Paradise*, both broadcast in 1960. Among the many sequences from this trip to New Guinea was footage of a sing-sing: a gathering of tribes where it is traditional to meet and gossip and share dance and music. And each person there was dressed more or less from head to foot in the feathers of birds of paradise. It was a shattering sight even back then and even in black and white. Though Attenborough didn't labour the point (he never labours a point), it was obvious that this is a frighteningly destructive way to live: that the more greatly you love the feathers the sooner you will run out of birds. I'm not sure I understood the word 'unsustainable' when I watched Attenborough's programmes at the age of nine, and I'm pretty certain the word wasn't much used in such a context. But I understood the concept of unsustainable lifestyles at that moment.

Though I could hardly disapprove of the idea that feathers are wonderful things. When we went to the countryside I always swooped with delight on any pheasant tail feather I found, so that I might be Robin Hood, who came to Sherwood Forest with a feather in his hat. I understood the idea very well indeed. But the practice of collecting feathers from living birds—in such lavish profusion, in such glorious excess— was clearly a kind of disaster.

And so it was, but it took the developed world to show the rest how to do the job properly.

Feathers are beautiful in themselves and so they give beauty to those who wear them. Certainly Ginger Rogers thought so when she was starring alongside Fred Astaire in their 1936 film *Top Hat*. One of the biggest numbers in the film is 'Cheek to Cheek', in which Fred makes his second attempt to win Ginger and they both spiral off into a lavish dance sequence. ('Everything Fred did I did backwards in high heels,' Rogers said on more than one occasion, when it was suggested that Astaire was the great dancer who was able to carry his less accomplished partner.)

left:
OSTRICH.
There's a famous
image in the *Egyptian
Book of the Dead* in
which the jackal-headed
god Anubis weighs
a human soul against
an ostrich feather.

Rogers was determined to wear a feathered dress for this dance, a blue creation decorated with 'myriads of ostrich feathers' as she described it. Both Astaire and the director, Mark Sandrich, wanted her to wear a white dress from a previous film. This provoked a stand-off. Neither side would yield. Rogers threatened to walk off set and then, still worse, she summoned her formidable mother.

Eventually they filmed the sequence in the feathered dress, which was fine except that feathers kept flying off the dress. 'It was like a chicken being attacked by a coyote,' Astaire said later. Astaire was furious. Hard words were spoken. But Rogers still wanted to wear the feathered dress. So there was an all-night session for wardrobe, the workers sewing on each individual feather to make sure that it stayed on the frock. The sequence was filmed again the next day, and this time things went better. Though if you look hard you can still see a few feathers flying about.

Astaire gave Rogers a gold feather for her charm bracelet as an apology, with the note: 'Dear Feathers, I love ya—Fred'. And she carried the nickname 'Feathers' forever after. It was a great sequence, and Astaire spoofed it in his 1948 film *Wedding Bells*, in which he starred with Judy Garland.

The moral of this story is that feathers matter to humans. Perhaps because, being implements of flight, they bring us closer to heaven.

> *Heaven!*
> *I'm in heaven,*
> *And my heart beats so that I can hardly speak,*
> *And I seem to find the happiness I seek*
> *When we're out together dancing cheek to cheek.*

*If you want
to climb Everest,
you're best advised
to take a lot
of down feathers*

Down feathers are the best insulators on the planet; we still haven't come up with anything better. We use them in pillows and duvets, we use them for the finest-quality sleeping bags and we use them in clothing. If you want to climb Everest, you're best advised to take a lot of down feathers. It's their lack of barbicels (see page 181) that enables them to trap the air so efficiently, giving them an unbeatable combination of lightness and warmth.

When you wear feathers you appreciate at first hand their drawback: once they get wet they lose all of their insulating properties more or less instantly. All the more reason, then, to wonder at the waterproofing capabilities of the outer feathers of most birds.

Most of us have taken a brisk winter walk around some lake or estuary and taken for granted the fact that ducks and gulls are sitting on the water—water we wouldn't so much as paddle in. We are kept warm by our down garments and dry by the waterproofing techniques we have invented. And we would hesitate to dip a toe into the water, while there is a raft of ducks bobbing about on the surface, apparently dozing at their ease, while all the time enjoying a temperature a few degrees below that at which you'd want your lager served in the summer.

Those outer feathers, oiled by the substances secreted from the preen gland and attended to by the bird with almost dandified care, are the secret: and beneath this impermeable overcoat, the marvellous insulation of down. Down has 'loft': the ability to expand from a compressed state and then trap air.

We still use the term eiderdown for a bed covering. Eiders are northern-hemisphere ducks—holders of the straight-and-level-flight speed record, remember (see page 17)—with a marked taste for sitting on very cold water. They line their nests with down they pluck from their breasts. This has been traditionally collected by humans, and the practice continues in Iceland, Siberia and Scandinavia. It can be done with discretion, without harming the birds, though it must piss them off a bit.

But most of the down we use comes from domestic stock, and it's a by-product of the meat trade. About 70 per cent of it comes from China. There have been issues there with live plucking, which people from the West find unacceptable. It's a classic example of the complex cultural confusions that surround our feelings about non-human animals. I have seen people in Chinese markets plucking live quails, to local indifference and the distress of Westerners—many of whom have no problems at all with eating chicken from fast-food outlets. After all, very few people get to see what happens on a battery farm.

The first great revolution in the history of the conservation of wild animals came about because of feathers. It's probably not true to say that without the issue of feathers there would have been no conservation movement: but it was with feathers that it all began.

It's about fashion. It's about women's hats. In the late nineteenth century and early twentieth it was important for women to wear vast hats with flamboyant feathers. This was about beauty, as then perceived (remember that the fashions of today are always the laughing stock of the next generation), and, of course, about prestige. A copiously feathered hat was a loud and proud statement of personal and/or vicarious wealth.

It followed, then, that the plume trade was big business. Among the more prized feathers were the breeding plumes of egrets. These are non-functional feathers that the egrets grow on their backs at the crucial time of the year: gauzy and floaty and immensely desirable. They were usually given the fancy term 'aigrettes', as if putting the word into French took all the harm out of it. (In the same way, we eat beef, pork, mutton and veal rather than cattle, pigs, sheep and calves.) These feathers were also, inaccurately, called ospreys. Ospreys are birds of prey and we had a completely different rationale for driving ospreys towards extinction (see page 237).

As a result of the plume trade, snowy egrets, the most desired species, were almost wiped out in the United States. There was also a vigorous trade in the feathers of flamingos, roseate spoonbills, great egrets and peacocks. It's been estimated that more than five million birds were killed every year to keep the trade going.

But don't blame it all on the whims of women. Wild-caught plumes were part of British Army uniform until 1889, worn on the cap. The entire plume trade was all the more pernicious because the birds were mostly shot in spring, so that their nuptial plumes could be harvested. This meant that the birds couldn't breed, so there was no possibility whatsoever of the practice being sustainable. But who was to stop it? Feathers were quite literally worth their weight in gold.

Naturally and inevitably, the feathers of birds of paradise became part of the trade, which caused more harm to the birds than anything the indigenous peoples of New Guinea could do (see page 187). The developed world brought an industrial scale to the business: 80,000 birds of

It's been estimated that more than five million birds were killed every year to keep the trade going

paradise were killed and exported from New Guinea in 1900.

But something about this trade got under people's skin. It was perhaps the frivolity of it, the wantonness of it, the complete uselessness of it. There was a slow but growing backlash: the trade in feathers was seen as flamboyantly wasteful—almost gratuitously irresponsible. Perhaps for the first time there was a widespread realization that the wild world was not, in fact, a bottomless well, a self-replenishing horn of plenty, from which we could help ourselves any time we wanted.

There was a sudden understanding that the earth's resources were finite, and that—at least in this one reckless and heartlessly frivolous instance—we could do something about it. Out of the extreme silliness of the plume trade came one of the first shafts of sense. And so the conservation movement began.

10

The Eureka birds

N ot finches. Mockingbirds.

As a species we humans prefer legends to facts, and the legend of Charles Darwin states that when he was wandering about on the Galápagos Islands, he saw a flock of the unique finches that live there and cried, 'Eureka!' It wasn't like that at all. Darwin came back from his gap year on the *Beagle*—actually the *Beagle*'s circumnavigation took five years, from 1831 to 1836—with a treasure trove of specimens that included the famous birds, the group now known as Darwin's finches. But the fact is that he didn't take the trouble to make a note of which of the Galápagos Islands each finch came from. Which, knowing what we know now, was close to being a disaster.

But he did make this all-important note with the mockingbirds. He collected four different specimens from four different islands, and he recorded the island of origin on the label of each one. He did this because he was, despite his error with the finches, a damn good observer, one of the best there has ever been. There's no criticism of Darwin's methods involved here; you can't judge people with hindsight. All the same, the fact that he didn't pick up the island-to-island differences in the finches is a bit of a look-behindjer moment.

But he did notice the mockingbirds, and that was massively significant: perhaps the most important bit of birdwatching in the history of humanity. He noticed that they were different, and he also noticed that they were very similar to the mockingbird species he had observed in

previous page:
MOCKINGBIRD.
For caption, see page 199.

Chile… but their songs were different. Similar but different. Ha!

Darwin was a man of action while he was on the *Beagle*, but he was a thinker all his life. Even at this early stage he was pondering the big question, and making notes. There were times when he seemed to think through his pen. The discipline of putting words on paper seemed to force his mind to come up with coherent arguments and logical—if sometimes deeply alarming—conclusions.

He was cataloguing some of the Galápagos specimens on the *Beagle* when he scribbled down this thought:

> When I see these islands in sight of each other, and possessed of but a scanty stock of animals, tenanted by these birds but slightly different in structure and filling the same place in nature, I must suspect they are only varieties… If there is the slightest foundation for these remarks the zoology of these archipelagoes will be well worth examining; for such facts would undermine the stability of species.

Those last four words might as well have been written in fire.

Different varieties of the same species can mate and breed successfully, producing young that will be capable of doing the same. Creatures of different species can't. In other words, Darwin was already fiddling with the big question. Here was a rock the size of the moon, but Darwin was feeding it into the rock crusher of his mind.

The treasure trove of specimens went to different experts to be examined when Darwin and the *Beagle* returned. The birds went to John Gould, ornithologist and ornithological artist. To Darwin's immense

right: GALÁPAGOS FINCHES, drawn by the ornithological artist John Gould.

surprise he pronounced that many of the apparently different birds he had brought back from the Galápagos were all finches and quite closely related, despite their considerable superficial differences. He also declared that the mockingbirds were all members of different species. Not varieties: species. Yet closely related to each other and all of them closely related to the mockingbird from the mainland of South America. How could this be? It was an immensely suggestive discovery. One that could possibly undermine the stability of species.

Einstein's notebooks are gloriously incomprehensible to the layman. Leonardo's notebooks are heart-breakingly beautiful. But perhaps the finest of all such examples of thinking on paper is Darwin's Notebook B. He kept this in 1837, which is to say, a year after his return on the *Beagle*. His handwriting is not beautiful, certainly not in these notebook productions, which were put together only to communicate with himself, but it has a certain compelling quality.

He wasn't much of a hand at graphic representation, either. But he put together the most significant doodle in the history of science and the history of human ideas and human culture. In Notebook B he drew a tree (see page 320). It has half a dozen main branches, and twenty or so subsidiary ones branching off. One of these descends to the ground. It is a tree of life, nothing less. It shows how one species can produce another, or more than one other. It also shows that some lines will go extinct.

Above this he wrote two words in his decisive scrawl.

'I think.'

Yes, sure—but how? The line of thinking was daring, but not entirely revolutionary. Not on its own. Unless he could come up with a mechanism to show how the whole damn thing worked, this was just an amateur's bit of guesswork. It was meaningless.

The Eureka Moment didn't happen on Galápagos. It happened in England in 1838, two years after Darwin's return from his voyage on the *Beagle*. He had been pondering the nature of species, as prompted by the mockingbirds and the finches, and he was wondering about the meaning behind it all. And then he read Thomas Malthus.

Malthus, an English clergyman, wrote his *Essay on the Principle of Population* back in 1798. In this he pointed out that human population increases in line with the supply of food. He was concerned about the dangers of overpopulation.

Darwin thought about this and set off in his own direction. Wild animals produce a lot of young animals, and yet, all things being equal, their populations tend to remain stable. In other words, more young are produced than survive to adulthood to breed in their turn. So what separates the survivors from those who fall short? Could it be that those who survive do so because they are better at surviving? Do they survive because they have an edge? And, if so, could it be that they pass these traits for success on to their descendants? And if these descendants had a more exaggerated version of that trait for success, could that trait hand them a further advantage? And could they in turn pass that on to their descendants?

After the mockingbirds, pigeons are the most significant birds in Darwin's revolution

Here was revolution. Here was an idea that would be shouted down in fury at the first sight. So Darwin preferred not to risk it. He sat tight on the big idea for more than twenty years.

In 1855 Darwin turned to pigeons. After the mockingbirds, pigeons are the most significant birds in Darwin's revolution. He was the most meticulous of men, and had no objection to getting his hands dirty. He wasn't a gentleman amateur, a thinker from the rich man's Olympus who tossed his thoughts down to the world when he thought fit. He was a man who pursued everything to the limit. When he decided that he needed to add an impressive body of work on species to his CV, he devoted eight years of his life to the study of barnacles, and much of what he did remains relevant today.

So it was with pigeons. His idea that a species could alter over time was considered by some to be fanciful: but it was staring the world in the face in the form of every domestic animal that humans have ever bred. Darwin made a radical imaginative leap to encompass this: what human breeders could do by means of artificial selection, as he termed it, so the processes that shape the world could do by means of... natural selection.

'Get young pigeons.' In that year of 1855 he wrote this instruction to himself in his diary. He had, over the previous fifteen years, read a great deal about the breeding of domestic animals, and had acquired a vast amount of knowledge about pigs and poultry. And as always, the idea behind his revolution in thought was utterly simple, and could be demonstrated by any pigeon fancier in London.

Our domestic breeds are each descended from a single wild species. Most domestic ducks are descended from mallards, most domestic geese from greylag geese, and all chickens are descended from the same red junglefowl. They are the same species and share an ancestor, even though they can look radically different. Dogs can look even more various, and yet a Chihuahua and a Great Dane are members of the same species. I was once introduced to the progeny of a liaison between a female St Bernard and a male dachshund.

So it is with pigeons. Rock doves are the species from which all our domestic and our feral pigeons are descended. All the pouters, tumblers, fantails and runts are at base rock doves. So Darwin immersed himself in the lore and language of pigeons and became a pigeon fancier. He took this up only four years before *The Origin of Species* was launched into the world. He had a pigeon loft installed at Down House, now in the London suburb of Orpington, and filled it with many different varieties of pigeon.

He sought out the company of pigeon fanciers in London. This was a rum thing for a wealthy gentleman to do, but when Darwin was on the trail of an idea nothing would stop him. Social prejudice was never going to keep him out of the grog shops where fanciers gathered and talked bloodlines. He was fascinated by the tiny variations between individuals; variations that only an experienced fancier could see, but which led in subsequent generations to radically different birds—varieties that bring great joy to those who love them.

Darwin was one of those people. 'I love them to the extent that I cannot bear to kill and skeletonize them,' he wrote, and yet he did just that, and did so all the time. Love wasn't going to stop him getting to the truth. He killed them, as gently as he could, and then he boiled them so that he could examine their skeletons. He was fascinated by the fact that the hatchlings of most breeds look more or less identical, but grow into radically different adults. The conclusion to draw from this was both momentous and obvious.

People had mused on the mutability of species before, and done so in the manner of the thinker. None of them suspected that the answer to the problem lay in the farmyard, still less the pigeon loft. Part of Darwin's greatness lay in his ability to think simple: to see the bleeding obvious: and then to get down in the dirt and prove it.

His scientific friends were increasingly impressed by what he found. One reported that he appeared to have found the equivalent of fifteen different species and three different genera… but they were all in fact breeds of the same ancient rock dove.

If a human fancier could breed birds as different as fantails and pouters in the space of a few generations, what might nature do across the millennia? One gradual change could follow another, and so animals might change as drastically as—no, even more drastically than—Darwin's pigeons. The difference is that while the fancier can play God and select for the qualities he wants—like an ornate tail, like an immense throat, like the ability to loop the loop—in the natural world the pace is different. All the same, time and the often-shifting demands of the environment will do the same thing and make selections for certain advantageous traits. In the pigeon loft the birds ill-adapted to the fancier's whims are discarded, killed off, denied the chance to breed, while those that have the desired traits survive and breed and pass on the traits that the fancier desires. In the wild world, the creatures that fail to cope with the conditions they face fail to breed, while those that have an edge will do so—and pass on the edge they were fortunate enough to possess.

Darwin found the meaning of life in a flock of pigeons and a grog shop.

The Origin of Species was published in 1859, to a mixture of frenzied approval and appalled rejection. Thomas Huxley, who was to become Darwin's most vociferous supporter, declared: 'How extremely stupid not to have thought of that.' Benjamin Disraeli took another tack and spoke for the majority: 'The question is this—is man an ape or an angel? My Lord, I am on the side of the angels. I repudiate with indignation and abhorrence these new-fangled theories.'

Because that was the question that terrified the world beyond measure, and it's fair to say that it still does. *The Origin* doesn't actually touch on it, but the question hangs over every line. If you accept that evolution is a fact, then you accept that humans have ancestors in common with non-human animals. Here was the world's overwhelming question.

In the much-mythologized evolution debate held in the Univerity Museum in Oxford in June 1860, seven months after the publication of the *Origin*, Bishop Samuel Wilberforce, aka Soapy Sam, asked Huxley (accounts vary as to the words he used) whether he claimed descent from a monkey on his grandmother's or his grandfather's side. Huxley replied that he would not be ashamed to have a monkey for his grandfather, but

In the pigeon loft the birds ill-adapted to the fancier's whims are discarded, killed off, denied the chance to breed

he would be deeply 'ashamed to be connected with a man who used great gifts to obscure the truth'.

Darwin always knew that publishing his great theory would make him many enemies, which is one of the reasons why he delayed publication for so long. And, naturally, these enemies leapt on any potential weakness. Perhaps the greatest of these weaknesses was the lack of hard evidence. All the circumstantial evidence of pigeons and, for that matter, mockingbirds was open to speculation, but where was the hard, undeniable physical evidence? Where were the transitional forms that must exist if Darwin was right? Where, in short, were the missing links?

There's a slightly apologetic, almost peevish tone in some places in the *Origin*, where Darwin admits this weakness. He cites—quite justifiably—the paucity of the fossil record. The odds against any creature surviving as a fossil are millions and millions to one against. We're lucky to have the fossils we've got; there's really not much use in wishing there were more.

But that argument didn't really wash. There were no transitional fossils: therefore the theory of evolution by natural selection didn't stack up. Relief all round.

A bird was to end this complacency at a stroke.

In 1798 a German actor called Alois Senefelder invented lithography: the art of printing from stone. He was eager to use this new technique for the printing of theatrical literature. And he was able to use the right material to do so, because he lived in Solnhofen in Bavaria, in southern Germany, where you can find the best lithographic limestone in the world. Once the technique was established, the limestone was increasingly quarried and used for lithographic plates. It was treasured because of its ability to record very fine detail. It follows, then, that the fossils that occasionally turned up in this limestone were of dizzyingly fine quality.

In 1861, a Bavarian doctor was presented with a fossil from a quarry near Solnhofen—possibly by a patient eager to pay his bill in kind. It was a good bit of business for the good doctor; he sold it to the Natural History Museum in London for £700.

Is it a bird? Is it a dinosaur?

Yes.

*Here was
Archaeopteryx.
The Germans
called it the Urvogel:
the original bird*

Here was Archaeopteryx, a word coined from the Greek meaning first wing. The Germans called it the *Urvogel*: the original bird. It was unquestionably a reptile, like the dinosaurs, because it had jaws with sharp teeth, three-clawed fingers on each wing and a bone that ran the length of its tail. It was equally unquestionably a bird, because it had feathers. With the details on the incomparable Solnhofen stone you can make out the barbs, barbules and barbicels of the flight feathers.

So that was a facer for anti-Darwinists. This was unquestionably a transitional fossil. This was a link between birds and dinosaurs: a link that was no longer missing. And it came to light in London in 1861—just two years after the publication of the *Origin*. It was as if God had decided to lend a hand. As if God had decided to bestow his blessing on Darwin and all his works. As if God was pointing out the path to atheism.

Darwin was never one to crow, and besides, plenty of people continued—and continue to this day—to deny the facts that Darwin had set before us. But in the many subsequent editions of the *Origin*, he notes Archaeopteryx calmly and austerely and, though he puts this in other words, invites all his critics to put the *Urvogel* in their pipes and smoke it.

It was Sigmund Freud who declared that humankind had suffered three shattering blows to our pride and our self-image in the course of history. The first, he said, was Galileo's demonstration that the earth is not the centre of the universe or, for that matter, the solar system; it is just another orbiting rock. The second was Darwin's revelation that humans were a species of animal. The third, he modestly explained, was his own demonstration that humans were not even rational animals.

Well, it's just a theory, isn't it? In 1980, President Ronald Reagan of the United States was challenged on evolution. He replied: 'Well, it's a theory—it is a scientific theory only, and it has in recent years been challenged in the world of science and is not yet believed in the scientific community to be as infallible as it was once believed.'

The theory of evolution is indeed a scientific theory, Mr President. Like the theory of gravity, in fact. That is to say, it is the only existing explanation of life on earth that covers all the known facts. It's not something that you can believe or not believe according to choice. You can no more disbelieve in evolution than you can disbelieve in gravity. In other words, you can believe whatever you like, but you still won't fall off the surface of the earth, the earth won't stop revolving round the sun and, whether you like it or not, you share an ancestor with a chimpanzee.

That is an essential truth of the human condition: one of the greatest and most shattering truths ever revealed. And it was brought to us with the copious assistance of the birds: mockingbirds, pigeons and the *Urvogel*.

In a small Hampshire village there is a small museum. It is divided into two parts. The top floor is devoted to the life of Captain Lawrence Oates, the man whose self-sacrifice on Scott's expedition to the Antarctic has become one of the great archetypes of British heroism. 'I am just going outside and may be some time,' he famously said before walking from his tent into the blizzards of eternity. Here was a man who travelled to the furthest limit of the earth to look for the planet's secrets and discovered instead the limits of human endeavour.

The rest of the museum is given over to a man who hardly ever set foot outside the parish of Selborne. By living in this way he revealed a series of truths about the way that life on earth operates. He changed the way we humans understand the wild world, and he did so because he was mad about birds. Well, he was fascinated by all wildlife, but it seems to me that he was a birder before he was anything else. In fact, he can claim to be the world's first birdwatcher, so he invented a hobby—interest—obsession—passion—as well.

White was a brilliant noticer—he was a genius of the commonplace: the great master of the ordinary

His name, of course, was Gilbert White, and we've met him earlier in these pages as the pioneer of three distinct sciences (see page 101). He was curate of the parish of Selborne on four separate occasions, and during the many years he was there he studied the wildlife of the place. He wasn't the sort to push himself forward and write books, but he was a great writer of letters. He wrote of his observations to Daines Barrington, a fellow of the Royal Society, and to Thomas Pennant, a zoologist. These letters were put together and published in 1789 as *The Natural History and Antiquities of Selborne* and have never been out of print. And they were as revolutionary as anything else that happened in that year.

White was the prototype parson-naturalist, something that became a widespread species in the course of the following century. At one stage Darwin considered such a life for himself. White devoted himself to this one small patch of the world and by doing so revealed truths that some of the great travelling naturalists—like Darwin—were subsequently able to profit from.

White was the first naturalist who did almost all of his work with living creatures out in the wild. This seems a fairly obvious thing to do now, but back then it was a game-changer. It was a radically new approach and it was phenomenally difficult. In those days it was very hard to get a definite identification of a bird without shooting it; there were no precision-engineered binoculars to hand, no sound recordings, no field guides, no birding apps for his iPhone.

But White was a brilliant noticer. He had a mind that was able to retain subtle memories of sight and sound. Thus he was a genius of the commonplace: the great master of the ordinary.

Which brings us to the willow wren: a small bird not easy to see, more often heard. White wrote: 'I have now, past dispute, made out three distinct species of willow wren.' He realized, in short, that the willow warbler, the chiffchaff and the wood warbler are three different birds. They sing quite different songs, they keep themselves separate and they certainly don't mate and breed together. White was also the first person to recognize the harvest mouse and the noctule bat as separate species.

right:
WILLOW WARBLER.
The Reverend Gilbert White realized that the bird known as willow wren was in fact three separate species: the willow warbler; the chiffchaff; and the wood warbler.

These are small things in their way: small creatures that come from a small place, and their identification, while interesting enough in a small way, is hardly the stuff of revolution. Rather, they demonstrate the brilliance and the meticulousness of White as an observer, as a tuner-inner. He was able to understand what the place, what the landscape, means; how it shifts, how it changes, how it renews, how it *works*.

White was, perhaps more than anything else, fascinated by the changing seasons. As we have seen, he spent much of his life preoccupied by the question of what swallows do in the winter (see page 102). He also loved first appearances, first flowerings: the annual miracle of the world that swings from harrowing cold to life-giving warmth and back again. The science of phenology has its beginning here, one that has become increasingly relevant in recent years with the changing of the climate.

So the work of a curious parson—curious in both senses of the term—changed the way we think about the wild world. He taught the world to observe, and to understand. And more than anything else, he was inspired to do so by the birds.

Gilbert White was also a writer, a man who brought things to a wider audience. He was good at it too: his book remains a damn good read to this day. So he was the first person who brought the wild world to an audience that was unfamiliar with it. He brought nature to the people of cities, reminded them that there was all that wondrous stuff out there. As urban life slowly began to predominate over rural life and more and more people became town dwellers, White was there to remind them that another way of life existed.

So White was also the first wildlife presenter. His skills as a communicator taught people to value the wild world—people who wouldn't otherwise have been in contact with it.

You can put White in a long line: Darwin himself, H. W. Bates, Alfred Russel Wallace, Jean-Henri Fabre, Gerald Durrell, David Attenborough, Peter Scott, Bill Oddie, Chris Packham… Individuals who have opened people's eyes and ears and hearts and minds.

Who's the head bull-goose loony around here?

White was the first wildlife presenter. His skills as a communicator taught people to value the wild world

That's the question Randle McMurphy asks when he enters the asylum in Ken Kesey's novel *One Flew Over the Cuckoo's Nest*. He was asking who's in charge. Who's top of the dominance hierarchy? He wanted to know this so that he could depose him and put himself in charge, and he does so effortlessly.

In recent years the term 'alpha male' has been borrowed from the study of animal behaviour and widely—even recklessly—used to describe human males. The idea is that human social life is hierarchical and that it works by establishing who defers to whom. Like most casually appropriated scientific concepts, it's both helpful and unhelpful. Sometimes those who seem most eager to establish themselves as number one are shunned and avoided, in human and other societies.

We also use the term 'pecking order' (or peck order) to denote the same rough-and-ready concept of a hierarchy of status in human societies. And it all began with chickens: so this was yet another revelation that came from the birds.

The phrase 'pecking order' lives on, but the name of the man who first noted it is largely forgotten, at least in Britain. He was a Norwegian zoologist called Thorleif Schjelderup-Ebbe. He discovered the phenomenon in 1921 and, by doing so, inadvertently handed humanity a helpful tool in trying to understand itself. He noted that chickens don't fight to the death over every morsel of food—this would be a pretty inefficient way for any society to operate. Rather the top chicken—the bull-goose chook—helps herself.

This doesn't demonstrate the brutality of nature but the opposite. If you don't know who defers to whom, you're going to spend an awful lot of time discussing who goes first and who has first bite. And there's likely to be blood spilt. But if you have already worked out the order of precedence, life can continue serenely. It's obviously best for the top bird, the alpha bird, but it's also better for the number two, for the beta bird. You don't want to get pecked to bits every time you fancy a bit of food. You wait your turn—and give hell to the bird below you, hence the old image of the disgruntled worker bullied by his boss who goes home and 'kicks the cat'.

You can observe this behaviour in the wild. I have watched avocets in spring, sorting out partners and territories, and I witnessed one protracted dispute. At the end of it, the defeated bird flew straight across the lagoon and picked on a bird lower in the hierarchy and gave him a brief but effective bad time.

We humans have formalized this concept for centuries. The army is the most obvious example: even the most raffish of guerrilla armies operates on the principle that you do what the person in charge says without debating the matter. We operate a similar if slightly less rigid system in the workplace, and it's so effective—being as natural for us as it is for chickens—that the boss seldom needs to remind a subordinate, 'I'm not asking you. I'm telling you.' A hierarchy is a way of simplifying life and making it more efficient.

But if a hierarchy is to continue as an enduring system, it must be open to change. A dominant bird or person can be challenged, and must be capable of slapping down a challenge. It is accepted that a challenge is a relatively big deal, though; not something that happens every day of the week. It's a process that fascinates us, being part of our nature, and so we choose to dramatize it for our entertainment. And that process is called sport. Tennis, in particular, is about hierarchy, and the sport is based around a ranking system, with the proviso that the bull-goose player is regularly available for challenge: either to reassert dominance or to lose the place at the top.

It's also been observed in chickens (and in many other species) that a high place in the hierarchy can be inherited. The chick of a dominant female automatically gets a higher place in the hierarchy than the chick of a less dominant bird. Chickens have an aristocracy… the British laugh at this and then wonder why this is a country that has so often chosen to be governed by Old Etonians.

Chickens have taught us a great deal about ourselves and the way we choose to live.

Chickens have an aristocracy… the British laugh at this and then wonder why this is a country that has so often chosen to be governed by Old Etonians

A friend was explaining to me over a pint why an ex-girlfriend of his used to carry goose eggs in her bra. It was about imprinting, obviously. She was researching animal behaviour, and she wanted to be the first moving thing that the newly hatched goslings set eyes on. The bra trick was the infallible answer. And it all goes back to Konrad Lorenz.

Lorenz was a scientist born in Austria in the early years of the last century. Lorenz studied animal behaviour, but he also liked animals. That was unusual back then. In fact, many thought it close to a contradiction. Lorenz liked to keep pets, and did so from boyhood. He was particularly fond of members of the crow family, and successfully raised many of them from nestlings. They became his companions, and he got to understand the way they conduct themselves. The way they operate. The way they think.

As you'll see, we are moving on to dangerous territory here. We are shifting away from the safe and severe confines of the laboratory, in which everything can be measured, and moving towards something woollier, more vague, more complex—and more real. Animals are not truly themselves in a laboratory, and perhaps that's also true for the human species. Outside the lab there's a bit more scope.

Lorenz was particularly interested in hard-wired instincts, and how such things operate as part of a bird's evolutionary equipment. Instincts are part of everybody's survival package: involuntary, more or less uncontrollable responses that enable us to do the right thing. My most vivid experience of this was when I encountered a lion face to face in the Luangwa Valley in Zambia. My instinct—one I didn't know I possessed—told me to stand stock still and not to move a muscle. Had I fled, the lion would have chased me and I would have been overhauled in a couple of seconds. But because I stood still, the lion—not hunting, not even interested in food at that moment, thank God—decided to back down.

Lorenz uncovered the mystery of imprinting as an essential survival trait in greylag geese. He discovered that geese will treat the first

living thing they see as a parent, and will follow it, and will be forever after attracted to the same model. Geese don't spend much time on the nest; they are born good to go, and so they need to go in the right direction: the direction taken by the parent. When they grow older, the same imprinting will help them to select the right species of bird to hang out with, socialize with, feed with and roost with, and—in the fullness of time—mate with.

Lorenz won the Nobel Prize for his work in 1973, in company with Nikolaas Tinbergen and Karl von Frisch. This wasn't just a tribute to their work in the narrow field of imprinting. It was recognition of the fact that they had formulated a new branch of biology. That was ethology: the study of living creatures, what they do and why they do it. And it's better practised in the wild.

I mentioned earlier that Gilbert White was one of the first people to prefer the study of living animals to that of dead ones. Lorenz and his colleagues turned that into a formal and legitimate science, and established a new way of study. It's no coincidence that many of the commonly used terms in ethology are German, in tribute to the tradition that Lorenz established. It was a science that involved a certain amount of intuition, empathy—even that dreaded stuff, anthropomorphism, or using the human imagination to try to understand what is happening with the animal in question. We are limited by our humanity, just as birds are limited by their avianity, but it's a boundary that humans can cross with the help of observation and empathy.

So Lorenz also established the fact that birds feel emotions. This was something categorically denied by laboratory scientists, who refused to entertain the idea. Lorenz took the softer view, something close to an anti-science view, and by doing so advanced scientific knowledge further than humans had been able to before. His was a triumph of the imagination.

His biography is compromised and confused by his Nazi past. We like to make secular saints of people who have done great work, and the more so if they have been able to spread this work in a way that the rest of us can understand. Lorenz's books of popular science are still worth reading. He wrote *King Solomon's Ring*, using the old story that Solomon was able to understand what animals were saying thanks to a magic ring. His more challenging *On Aggression* is about aggressive behaviour as a survival mechanism, with the suggestion that this could be channelled and modified in humans. I spent an awful lot of my life as a sports writer and Lorenz is not wrong. I have found my own reading about evolution and ethology a vivid help in understanding the way sport works.

Lorenz was a member of the Nazi party from 1938 and, at least for a time, held views about race that, dubious during the Second World War, are anathema now. So let's not celebrate him for what he wasn't. But he was a man whose vision and understanding brought humans closer to non-human animals, and made it possible to reassess both.

Jays, like many birds, have a tradition of gift-giving as part of courtship and bonding

The jay observed the waxworm with desire. He assessed the situation. The worm was in a glass jar, and the jar was too narrow to admit his beak. The worm was floating in water at the bottom of the jar. So the jay filled the jar with pebbles until the worm floated close enough to the rim of the vase… and took his reward with relish. This is nothing less than one of Aesop's fables come to life.

I observed this behaviour when I went to Madingley, near Cambridge. This is the domain of Nickie Clayton, Professor of Comparative Cognition at Cambridge University. Here she keeps an aviary of corvids, crows and their relations. She was towering well over five feet tall in extravagant heels; she dances the Argentine tango for twelve hours every week and she is scientific adviser to the Rambert Dance Company. So she's quite something.

'Anything an ape can do, so can a crow,' she told me. 'Crows belong in the Clever Club along with apes, elephants and dolphins.' The work continues: problem-solving, learning, memory and planning for the future. Like humans, crows can anticipate difficulties and plan accordingly. One of Nicky's students was working on a project about whether male jays know what female jays want. Jays, like many birds, have a tradition of gift-giving as part of courtship and bonding. Just as a husband brings his wife a cup of tea in the morning rather than a pint of lager, so this student was working out if a male jay's gifts are what the female actually wants.

All this research is taking notions of intelligence in birds away from anecdote and fable and replacing such things with hard data. All of it seems to show that we have always been too jealous of our own intelligence. Throughout history we have begrudged the idea that we share it with any non-human animal—let alone a bird. But it seems that birds are not bird-brained after all. Time we wised up.

11

The way of the dodo

H

ow many deaths will it take till we know—that too many species have died? The answer to that one may well be blowing in the wind, but one certain fact is that it took us an awful long time to cotton on to this issue. Extinction has been a fact of life—and an accelerating fact of life—for as long as humans have been the dominant species on the planet. Say, five millennia. And yet it's only in very recent times that we have become aware of it.

I remember a scene in a comic novel, long out of print and out of fashion, but a girlhood favourite of my mother's. The book is *Half a Sovereign* by Ian Hay, a 1920s romp with a great gathering of disembodied spirits thrown in. In the climactic scene, members of the assembled band of spooks start to vanish, one by one, and the hero wonders how long it will take before the female villain notices. That's pretty much how it's been for non-human life during the last three or four centuries of human existence.

Birds have been vanishing from before our eyes. How many times must a man turn his head and pretend that he just doesn't see? Well, plenty of people prefer to carry on doing just that.

Birds are the one group of creatures that we're always aware of, so you'd have thought we'd notice that their numbers were declining or that a particular species was no longer around. But the notion that we have an extinction crisis on our hands has only been widely accepted for the past couple of decades.

previous page:
THE DODO.
For caption, see page 277.

The dodo was first noticed in the West in 1598. The last recorded sighting of a dodo was in 1662. Didn't take us long, then. But the whole business was for years regarded as more comic than tragic.

The dodo lived on the island of Mauritius, east of Madagascar, in the Indian Ocean. It was one of those island birds that evolved from a mainland ancestor back in prehistory, like the finches and mockingbirds of the Galápagos. Like the Galápagos cormorant, the dodo evolved flightless. No one told the blind forces of evolution that humans were coming and would change everything; that they would soon be shifting the goalposts.

The dodo's ancestor was a pigeon, though this was considered a ludicrous notion at one stage. Other contenders were put forward instead: ostrich, rail, albatross and vulture. But here was a giant flightless pigeon, a fruit-eater that thrived by walking about with a stately pigeon gait through the rich forests of the island.

The Dutch called it the *wahlghvogel*, which means 'tasteless bird'—not hard to work out the priorities of those that did the naming. It's suspected that the name we know came from a Dutch nickname for the bird. There are two contenders: *dodoor*, which means 'sluggard', or *doodars*, which means 'fat-arse', and which could be a reference to the great bustle of feathers the dodo carried on its back end. I prefer the second explanation: the dodos were unlikely to be sluggards. Avian anatomists believe that the strong legs would have made the dodos very effective runners when they needed to be.

They were impressive beasts, standing more than 1 metre (3 ft) tall. The biggest males—larger than the females—could weigh up to 21 kg (47 lbs). A lot of meat on them, then; even if you reckoned it was tasteless, these birds carried a lot of protein. They were also largely unafraid of humans, since they hadn't evolved as a prey species.

In recent years we've come to accept a mythical version of the dodo's extinction, in which the kind, confiding, sweet-natured dodos were ruthlessly rounded up and eaten by callous humans—people who gave no

thought whatsoever to what they were up to. They were wiped out by a casual sideswipe from a species with other priorities on its mind.

It's now largely agreed that this wasn't the full story. It wasn't just that humans hunted dodos as if there was an inexhaustible supply of them. It was also that humans radically changed the island of Mauritius. They cut down vast tracts of forest for timber and to clear land for farming, and forest was the dodo's home. They also introduced non-native animals to the island: and the pigs and dogs, it's reckoned, devoured the dodos' nests until there was none left.

So it wasn't just human callousness that did for the dodo. It was also human carelessness. All one for the dodo.

The dodo was established as an icon of loss after a member of the species appeared as a character in *Alice's Adventures in Wonderland*, which was published in 1865. The author was Lewis Carroll, whose real name was Charles Dodgson. The Dodo is a self-portrait, a joke against himself: he was a stutterer who habitually introduced himself as Do-Do-Dodgson.

The dodo's appearance in the book two centuries after its extinction, along with John Tenniel's brilliant illustrations, brought it into the mainstream, giving rise to such expressions as 'dead as a dodo' and 'gone the way of the dodo'. It seems there was little self-reproof implied in such phrases. The dodo was largely seen as a bird that was too stupid to live—one that probably deserved to be wiped out. The new notions of Darwinism and survival of the fittest helped to cement the idea.

It's far from unusual for islands to develop unique species, usually called 'endemics'. It's part of the pattern of evolution and, as we have seen, part of the key that enabled Darwin to unlock its secrets. These island species tend to have small populations, as befits a small place. It follows, then, that it doesn't take much to wipe them out. Other Mauritian birds that have gone the way of the dodo include red rail, broad-billed parrot, Mascarene grey parakeet, Mauritius blue pigeon, Mauritius owl, Mascarene coot, Mauritius shelduck, Mauritius duck and Mauritius night heron. The dodo was not alone. It was just part of the holocaust that hit Mauritius when humans came to call. The Mauritius kestrel, which we met earlier, was unusually lucky.

Island species tend to have small populations, as befits a small place. It follows, then, that it doesn't take much to wipe them out

The great auk is remarkable: a law was passed for its protection. It didn't do much good. As we have seen (see page 91), the bird went extinct in the middle of the nineteenth century, but at least its passing was accompanied by a feeling that this was not entirely a good thing, or even an unavoidable thing.

The great auks were essentially birds of the open ocean that came ashore to breed. They were found off the coast of Canada, Greenland, Iceland, the Faroes, Norway, Ireland and Britain. They looked pretty similar to penguins and they evolved—via a quite separate route—for a very similar lifestyle. So here is a fine example of what's called convergent evolution: the same solution reached by quite different routes.

Great auks were dark on top and pale underneath, like many other fishing birds. From the point of view of a white-tailed eagle the auk was as dark as the surface of the ocean, but to an orca approaching from underneath, the bird was pale as the sky. This pallor also helped to conceal them from the fish they fed on. Great auks stood up to 85 cm (nearly 3 ft) high, and weighed up to 5 kg (11 lbs). Substantial things, then.

And like a penguin they were flightless, or, if you prefer, they flew underwater rather than through the air, on wings 15 cm (6 in) in length. They had a great heavy hooked beak, which was a very effective tool for fishing. It's been suggested that these birds hunted co-operatively, working a shoal in numbers. They did their breeding on rocky ocean islands, where they could scramble above the splash zone. The problem was that they were useful to humans in all kinds of ways and being flightless, they were—not quite literally—sitting ducks.

They could be eaten, of course. Great auk bones have been found at Neanderthal sites around 100,000 years old, so it's an ancient tradition. They also made a handy bit of bait for people going after substantial fish. But the clincher was their down. Birds that have evolved for a life on cold water need the best insulation, like the eiders already mentioned. The down from great auks was almost as desirable, and so the birds became a target species.

A law was passed in Britain in 1794 that made it illegal to kill great auks for their feathers, and people who did so could be punished with a public flogging. It was still legal to catch them and use them for bait, so you could argue about the thoroughness of these legal restraints. But here at least was a groping historical notion that the conservation of wildlife was a public concern and that the state had a role to play.

A good precedent, then—an important precedent—but it didn't do

much for the great auk. It was reckoned that the European population of great auks was seriously compromised by the sixteenth century, and it continued to decline despite the law in Britain. It's possible that the Little Ice Age—the period of very cold winters that lasted roughly from 1300 to 1870—affected the great auks by making their breeding colonies accessible to polar bears across frozen seas.

All the same, humans played their part in the extinction of the great auk, and they did so with élan. As the decline began to look irreversible a colony of about fifty birds was discovered on the small island of Eldey, off Iceland. This, alas, provoked a feeding frenzy from museums and private collectors, all of whom wanted a great auk in their collection before the world ran out. The last two great auks were killed on Eldey in 1844; there is an unconfirmed sighting from 1852.

So the world had progressed. Here was another extinction, yes, but at least it was an extinction that people tried to prevent.

If time travel were possible, I'd take my Tardis to New Zealand a few years before 1280. This is the ultimate destination in time and space for anyone with birding in the blood. New Zealand was the kingdom of the birds and it remained so until the arrival of humans in the thirteenth century.

New Zealand was isolated from all other land 65 million years ago. This is a monstrous amount of time for evolution to work in. And here evolution took the most extraordinary direction, because on this comparatively enormous landmass there were no mammals at all, apart from those that could get there by flying. The only New Zealand mammals were bats: three species of them, now reduced to two.

So here was a vast area of land with birds as the dominant large life forms. Naturally (once again, please notice the choice of words here) they evolved into endless different endemic forms. And they did so in a fashion that is quite remarkable, filling ecological niches that would normally be filled by mammals. The classic example is the kiwi: a bird that has adopted the mammalian lifestyle so effectively that it looks more like a mammal than a bird. It operates at night, it skulks about on the ground because it can't fly, it operates mainly by smell, being the only bird with nostrils at the tip rather than the base of its beak, and its feathers look like fur. They lack the hooks that keep feathers tight and feather-like.

Kiwis survive today, of course—five species of them, all under different degrees of vulnerability. But the great glory of New Zealand birds is gone. This was once the land of the bird; Captain Cook reported that when he landed in 1770 he was deafened by birdsong.

To walk those islands in pre-human times would have been to enter what was almost literally a different world—a place of evolutionary possibilities quite unlike our own. New Zealand is the country of 'what if'. What if mammals had failed to survive the K–T* extinction and the meteor impact that did for the dinosaurs 65 millions years ago? What would the world be like? The New Zealand Tardis would provide the answer, telling us not just the scientific details about which species existed and how they lived, but what the place was *like*. What would it feel like to

* K–T stands for Cretaceous–Tertiary.

be a mammalian, a human intruder, in the country of the birds?

That is what the Maoris experienced when they first arrived in New Zealand in around 1280. They would have had to deal, for a start, with birds several metres tall. There were nine species of moa, and the biggest of them stood up to 3.6 metres (12 ft) in height, though some avian anatomists reckon it carried its neck low and forward, in the manner of a kiwi. It was still a vastly imposing beast—and yet it had a predator.

Haast's eagle is reckoned to be the biggest eagle that ever flapped a wing. The top weight for one of today's eagles is 9 kg (20 lbs); it's reckoned that Haast's eagles could reach 15 kg (33 lbs). Their wings were relatively short, in the manner of a sparrowhawk built on a nightmarish scale. It is surmised that the bird was able to manoeuvre through dense scrub and launch ambush attacks on the plant-eating giant moas.

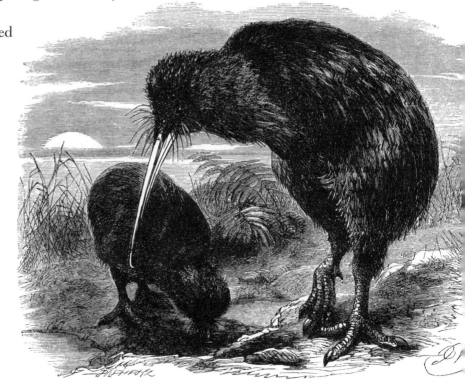

The Maori pioneers would have had to deal with giant eagles used to taking two-legged prey that stood six feet and more from the ground. Did they try their luck against humans too? Hard to believe they didn't. And did they succeed? Only the Tardis will tell you, though Maori legends tell of a giant bird that killed people. And remember that a golden eagle, though much smaller than the Haast's, is perfectly capable of killing a young deer.

Kakapos aren't much good at thriving. But they are, it seems, capable of hanging on

But it wasn't just these superstars that disappeared with the loss of the land of birds. It's been estimated that the pre-human New Zealand birds included 131 species that lived on or near the land, and 93 of them were found nowhere else on earth. Of these, 43 are now extinct, 15 of them since 1840. They've all gone for the usual reasons: habitat destruction, especially the felling of forests, and predation by introduced mammals. Once again humans radically shifted the goalposts of evolution and once again the birds lost.

Think of them: adzebill, laughing owl, long-billed wren. Eight species lost from Chatham Island (400 miles south-east of North Island) have Chatham in their common names, including a raven and a penguin. There were nine more whose common name bears the prefix 'New Zealand': the New Zealand bittern, coot, goose, musk duck, owlet-nightjar, quail, raven, stiff-tailed duck and swan.

All gone. And with them a world.

There have been some genuinely heroic efforts to save what's left of the great New Zealand avifauna. We humans are capable of learning and, for that matter, capable of all kinds of heroism. There are fine tales to be told about the rescue of the takahe, a flightless rail a bit like a moorhen; of the kiwis; and, perhaps most memorably, of the flightless parrot, the kakapo. This utterly improbable bird was wonderfully described by Douglas Adams, author of the *Hitchhiker* trilogy, in his book *Last Chance to See*: 'The kakapo is a bird out of time. If you look one in its large, round, greeny-brown face, it has a look of serenely innocent incomprehension that makes you want to hug it and tell it that everything will be all right, though you know that it probably will not be.'

That was written more than twenty years ago, and yet we still have a few kakapos, much cherished. I'd like to say thriving, but kakapos aren't much good at thriving. But they are, it seems, capable of hanging on, thanks to a little learning of the lessons of the past, plus the required epic quantities of heroism. I don't want to come across as too species-hating here. Humans are capable of doing some wonderful things: and looking after kakapos is a fine example of what people can do when they refuse to let concepts like 'impossible' stop them from doing the right thing.

We touched earlier on the Victorian slaughter of birds of prey, and we need to do so again, though briefly. All I shall do here is quote from a letter to *The Times*, from November 1877. The correspondent, F. O. Morris, concludes:

Only the other day a cousin of mine in the North Riding sent me, quite casually, the following list of birds of prey destroyed on one estate alone—Glengarry, in Scotland—in three years—namely, between 1838 and 1840. No wonder that some of them are totally extinct now, and others all but so, for the same work of destruction has been carried on more or less ever since. This is the list by Mr A. E. Knox. Here it is:

Twenty-seven white-tailed eagles, 15 golden eagles, 18 ospreys or fishing eagles, 98 blue hawks, 275 kites, 5 marsh harriers, 63 goshawks, 7 orange-legged falcons, 11 hobby hawks, 285 common buzzards, 371 rough-legged buzzards, 5 horny buzzards, 462 kestrels, 78 muleen hawks, 83 hen harriers, 9 ash-coloured hawks, 6 ger-falcons, 1,437 hooded or carrion crows, 475 ravens, 35 horned owls, 71 common fern owls, 3 golden owl, barn or white owl, comparatively rare in Scotland, and 8 magpies. On this occasion I have omitted the quadrupeds, which figured equally in this black list.

And for what purpose is all this fell slaughter? With your obliging leave, I will say something on this part of the subject in a future letter.

I am, Sir,
F. O. Morris

Good on yer, F. O. Some people were beginning to realize that the idea of humans fighting an underdog battle against nature was about a thousand years out of date. We needed a new take on the subject. This has spread considerably in the last century and a half—but it still has an awful long way to go.

To tidy things up, the blue hawk is the sparrowhawk, the orange-legged falcon is the red-legged falcon, the horny buzzard is a honey buzzard, muleen hawks are merlins, the ash-coloured hawk is Montagu's harrier, the ger-falcon is a gyrfalcon, the horned owl is a long-eared owl and the fern owl, bewilderingly, is the night-jar—a bird that eats insects.

The white-tailed eagle went extinct in Britain; the marsh harrier was reduced to a single pair; red kites became extinct in England but hung on in Wales; ospreys went extinct as breeding birds in Britain. All of these have come back, in some cases in decent numbers. Here is a tribute to a great deal of seriously hefty conservation work, so once again there is good news to be found in a chapter that can't help but be somewhat gloomy. It's not easy putting the toothpaste back into the tube, but sometimes it can be done. It's a question of having the will.

above:
THE GYRFALCON
is a rare winter visitor to Britain and Ireland.

12

The fightback begins

I t takes a lot to turn around any of the accepted truths of a society. The longer these truths have been accepted, the harder it is. When a belief has been part of the way we think for years, for decades, even for centuries, the people who question them need both vision and courage. When that belief has been part of the human condition for a million years or more, you need someone truly exceptional to take it on. Yet very few people know about Emily Williamson from Didsbury in Manchester.

As we have observed again and again in these pages, the prevailing view of humankind for generation after generation was that nature was a bottomless well, from which we could help ourselves eternally in a world without end. There was no concept of nature as a finite resource, no suggestion that humans had responsibility for any species beyond our own. (For that matter, the idea that we had any kind of duty towards any race of humanity beyond our own was, until comparatively recently, considered a dubious proposition at best. Slavery and genocide are part of the inheritance of the developed world.)

We took from nature. Plenty more where that came from. And if that process of taking should end in extinction, as it did with the dodo and the great auk and many others, well, it was a world owned and governed by humanity and if there were living things that didn't like it or couldn't stand the pace, we were better off without them.

Emily disagreed.

In particular she disagreed with the fashionable exploitation of birds for female garments. Horrifying figures from the feather trade have already been quoted and there is no need to go through them again. But Williamson hated all this. She hated the idea that the frivolities of fashion and vanity required mass death. So she decided to do something about it. Thus one of the world's largest and most significant conservation organizations was formed, one that has gone soaring way beyond the original idea of its founder.

Williamson founded the Plumage League in 1889. All of its members were women who thought that the international trade in birds' feathers was abhorrent and the fashion for wearing them obscene. They went on to amalgamate with the Fur and Feather League, which was based in Croydon, and so they became the Society for the Protection of Birds. The Society had but two rules:

> That members shall discourage the wanton destruction of birds, and
> interest themselves generally in their protection.
> That lady members shall refrain from wearing the feathers of any bird
> not killed for purposes of food, the ostrich only excepted.

It was a campaign that attracted the attention of women of high social standing; the first president was Winifred Cavendish-Bentinck, Countess of Portland. In 1904, the society was given its royal charter, just fifteen years after the amalgamation. Thus the Royal Society for the Protection of Birds—the RSPB—began. It now has more than one million members. In April 2011 it managed 141,830 hectares of land for nature conservation, of which 57 per cent was owned outright, with the rest under lease or management agreement.

Charles Rothschild was the second son of the 1st Baron Rothschild, so he was a man of wealth, position and influence. He was also nuts about nature. He was no dilettante: his work on fleas is relevant today. He was the first to describe the vector for bubonic plague, the Oriental rat flea, which he named *Xenopsylla cheopis*.

Nor was he just a lab man. He loved to be out in the real thing, out in nature, his eyes, ears and nose tingling with the wild world. He met his wife when butterfly-hunting in Hungary. He bought Woodwalton Fen, now in Cambridgeshire, one of the last chunks of fen that survived the great draining, and there he built himself a sort of safari bungalow in the Hungarian style. You can visit the place to this day. I have even had a picnic at Rothschild's table inside the bungalow.

It's instructive to note here that many of the early naturalists looked for plants and insects more than anything else. Birdwatching wasn't yet an accessible passion, mostly because there were no easily portable binoculars, still less easily affordable ones. Good naturalists were familiar with bird calls and knew what they were looking at, but that easy intimacy with birds that modern birders experience every time we put our binoculars to our eyes is a comparatively recent thing.

Rothschild loved the wild world. He stayed at his bungalow whenever his commitments to business and public life allowed. There is a wonderful picture of him on safari for Hungarian butterflies, with a stub of cigar in his mouth and a truly enviable hat.

Like Emily Williamson before him, he felt the shadow of loss. He was deeply aware of the fragile nature of the wild world that he loved. And being a person of means and influence he too set out to do something about it. But it wasn't the vulnerability of species that troubled him; it was the vulnerability of places. He was concerned that the spread of towns would swallow up the countryside. He could see that a rising population and increasing industrialization were eating up wild spaces, and that beautiful and treasured places, places as rich in wildlife as Rothschild was rich in money, were being destroyed, or were in great danger of being destroyed.

In 1912 Rothschild founded the Society for the Promotion of Nature Reserves (SPNR). Four years later he and his fellow members had completed a list of sites that they considered 'worthy of preservation'. It was a funky old list, idiosyncratic, mostly bringing in the members' favourite spots. They specifically targeted the places that had an uncertain future: that was the whole point—not just worthy of preservation but in urgent need of it.

In all they listed 284 sites in Britain, including Ireland, which was then considered to be the same thing. There was an instinctive bias towards places that were valuable for their plants and their invertebrates, especially butterflies, but there were plenty of birds lurking about on the list as well.

These included the little-tern colony near Aldeburgh in Suffolk, and Bass Rock, perhaps the most spectacular wildlife site in Britain. You can see it from the shore—also from the golf course—at North Berwick in Scotland. A boat trip, these days easily available from the Scottish Seabird Centre, will take you on a visit to the place, allowing you to join up with 150,000 or so gannets: birds with six-foot wingspans that

below:
COMMON TERNS. Light and graceful in flight, their forked tails have earned them the nickname 'sea swallow'.

dive into the sea for fish from heights of 30 metres (100 ft).

It was 1916 before the list was complete. At that time Britain had other things on its mind than the revolutionary notion of protecting wildlife. And so this outstanding idea got lost in the fog of war. In 1923 the SPNR lost its leader when Rothschild, a man haunted by depression and illness for much of his life, killed himself. It seemed that a great idea and a great man had both come to nothing.

You'd have thought that the next world war would only have made the situation worse. But during the Second World War the government began to make plans for the peace while the war was still going on. The SPNR established a Nature Reserves Investigation Committee in 1943, and their findings became government policy in 1947. Perhaps they wanted to wage war a little better the second time around. All the same, it was a monstrous act of faith. And it was here that the SPNR, an organization still holding on despite lack of funds, suddenly found itself in a position of real influence. They were the people on the spot, they knew what to do; they were too good to waste. They had been preparing for this moment most of their lives and they seized it with delight. Thus wildlife conservation slowly became an accepted part of national life. The SPNR went on to become the central organization for the network of county wildlife trusts that now cover England, Wales, Scotland, the Isle of Man and Alderney, and it is today known as the Wildlife Trusts. The county wildlife trusts collectively manage more than 2,300 reserves.

Roosevelt famously played truant with the British foreign secretary, Lord Grey, and the two of them had a long day's birding in the New Forest

The Americans were a bit quicker off the mark. So, it has to be said, were most nations. Britain set up its first national park in 1951. The United States had its own and the world's first national park as long ago as 1872: Yellowstone.

They had already accepted the idea of conservation as an aspect of national policy when a skinny, bespectacled New Yorker went to North Dakota in the 1880s. He really got it bad. He loved it all with a passion that was to last all his life. He learned to ride western style, to rope cattle and to shoot. But it wasn't just the way of life and the adventure that got to him; it was also the landscape and the interaction with non-human life. He felt that these places were precious—and also fragile. Perhaps you needed to come from a city—perhaps you needed to come from New York—to understand that. The wild places could no longer be taken for granted, and yet they were an essential part of America and the world. And he wanted to do something about it. His name was Theodore Roosevelt.

When he looked back over his presidency, which was between 1901 and 1909, he was inclined to look upon his conservation work as his greatest achievement. It was probably his most lasting one. On his watch five new national parks were created, and also fifty-one bird reserves. In all, 36,000 square miles (930,000 sq km) of land were given protected status, and federal protection was given to land and wildlife. Roosevelt was a great birder, and on a visit to England after his presidency he famously played truant with the British foreign secretary, Lord Grey, a fellow birder, and the two of them had a long day's birding in the New Forest.

The difference in awareness is shocking. In 1916, as we have seen, Charles Rothschild and the SPNR offered the list of sites 'worthy of preservation' to a sceptical nation and an uninterested government. In that same year, under the presidency of Woodrow Wilson, the United States passed the Organic Act. This founded the National Park Service and separated the national parks from the commercial demands of forestry. There it was, spelt out in the statute book: the fact that nature had a value *for itself*.

Thus America had laws in place to protect nature an entire generation before Britain got round to the same thing. In 2009, a video and accompanying book on the national parks was released under the title *America's Best Idea*.

And the world's.

In lowland Britain we expect the farmed landscape to be friendly to people and to wildlife

It's instructive to note that the British and the American ideas of how to look after the countryside are radically different. In lowland Britain we expect the farmed landscape to be friendly to people and to wildlife. We expect to be able to go for a nice walk and to see and hear the birds as we do so. We believe that the countryside should be beautiful: that it has a function both for the production of food and for the pleasure of humans, whether they are residents or visitors. That means the countryside must also support wildlife. It must echo with the sound of birds.

This idea of double value is alien to American thinking. There nature and agriculture are seen as an either/or. I drove through Nebraska on my way to Badlands National Park. Nebraska makes no attempt to be anything other than a factory for cereals, soy and cows. There is no thought of softening the purely functional aspect of this landscape. If you want nature, you can keep on driving north until you cross into South Dakota and get to the park—which is fabulous on a scale impossible to imagine in lowland Britain. This is a horizon-to-horizon wilderness. Of course, it helps that a great deal of the place is impossible to farm.

The idea that wild places and wild species needed protection took time to be universally accepted, even in the United States. In 1886, the editor of a magazine called *Forest and Stream* was the promisingly named George Bird Grinnell. He was appalled by the random killing of birds that went on, and so he founded the National Audubon Society in order to do something about it. The society was named for the great bird illustrator, John James Audubon. It was a great success, and remains so to this day. It had 39,000 members within a year and now has more than 500 local chapters.

It has always been my practice, when in the United States, to make contact with the local Audubon Society and ask for advice on where I might spend a day or even half a day playing truant (like Roosevelt and Grey) and going birding. I have had some memorable days as a result of this, and certainly the best was when I was covering the Ryder Cup golf event at Oakland Hills in Michigan in 2004. I was invited to join a party of retired ladies and after a very satisfactory morning involving sandhill crane, cedar waxwing and bald eagle, we went up to the Great Lakes in the afternoon for one of the greatest things I have seen, even though the competition for that accolade is pretty intense. In the space of two hours we saw 20,000 hawks migrating south, funnelling together over a land bridge between the lakes.

In the early days people were not always convinced that conservation mattered—even those who should know better. In 1902, Charles B. Cory, president of the American Ornithologists' Union (AOU), refused to attend a meeting of the District of Columbia Audubon Society on the grounds that 'I don't protect birds. I kill them.'

The notion that collecting birds for science was more important than looking after them and the places where they live is one that has died hard in certain quarters, most particularly America. There have been controversies in the current century—something that seems impossible to believe—with AOU expeditions shooting endangered birds because they thought it mattered for science, abiding by the ancient and out-moded doctrine, 'What's hit is history, what's missed is mystery.' The argument against this is clear enough, however important the science. Identification of a species can be confirmed with digital photography, which, in connection with a telescope, is capable of producing pin-sharp images across colossal distances. It's also possible to get enough stuff for DNA analysis without killing the bird. Droppings or a single feather will do the trick.

In the space of two hours we saw 20,000 hawks migrating south, funnelling together over a land bridge between the lakes

Conservation has always been plain common sense: and yet it has always been shockingly controversial. The idea that we should compromise the way we live or limit the ways in which we make money just to look after nature is still by no means universally accepted. The things that have become orthodoxy now were in their time most bitterly contested, with an aggression that looks to modern eyes to be deranged, and with a degree of misrepresentation that is clearly wilful.

John Keats wrote a poem called *La Belle Dame Sans Merci*, which is probably better not translated as 'The Beautiful Woman Who Never Said Thank You'. It begins:

> *Oh what can ail thee, knight-at-arms,*
> *Alone and palely loitering?*
> *The sedge is wither'd from the lake,*
> *And no birds sing.*

A writer and marine biologist thought that these lines could be adapted to make a nice chapter title for her current book. She was later persuaded to make it the title of the whole book. And so in 1962 in the United States Rachel Carson published *Silent Spring*. Sir David Attenborough described it as the most influential book since *The Origin of Species*.

The subject of the book was the detrimental effect of pesticides on the environment, but (like the *Origin*) it had implications far beyond its subject. *Silent Spring* is about the way that humans are destroying the natural world. The truth was there to be read, soberly and scientifically argued.

The title alone was a work of genius, for it brought home to people in just two words what was happening. The idea that spring might arrive without the song of birds is an obscenity. It's strange to think that the book prompted passionate opposition, but people declared that the aerial spraying of vast tracts of countryside with a mixture of DDT and fuel oil was an unambiguously good thing.

The chemical industry went to war over the matter. Carson was wilfully misrepresented; it was claimed that she said that all spraying of any kind had to be stopped, so clearly she was mad. One scientist said that if her advice was followed 'we would return to the dark ages… diseases and vermin would once again inherit the earth'. There are suggestions that ex-President Eisenhower said Carson was probably a communist, and the fact that she was unmarried made that all the more likely.

A letter published in the *New Yorker* went even further: 'Miss Rachel Carson's reference to the selfishness of insecticide manufacturers probably reflects her communist sympathies, like a lot of our writers these days. We can live without birds and animals, but, as the current market slump shows, we cannot live without business. As for insects, isn't it just like a woman to be scared to death of a few little bugs! As long as we have the H-bomb everything will be OK.'

You can't argue with facts like that.

Carson had made clear, for the first time, that development and scientific advance was not an unambiguously good thing

But despite and because of the opposition, *Silent Spring* conquered the world. Carson had made clear, for the first time, that development and scientific advance was not an unambiguously good thing. It was also obvious that big business existed to make money rather than improve the lot of the human race, and for this reason, business interests were not entirely to be trusted. It was a radical re-think on several levels.

But its deepest impact was not on government or business or law; nor, for that matter, was it on birdwatchers and naturalists. It was on ordinary people, who were appalled by the idea of living in a land where birds failed to greet the spring, and that there were people and organizations that were prepared to let that happen.

From this point on, conservation became an aspect of democracy. It's been said that democracy is not about voting or rights; it's about thousands and millions of people saying with one breath: 'They can't do that!'

That was the response to *Silent Spring*. It remains the bedrock of the environmental movement.

I remember seeing a buzzard when I was on a family holiday in Cornwall. Well, two of them, in fact—a double miracle. I was eleven at the time. I think it was my father who spotted them; anyway, I've always given him the credit over the years. It was quite astonishing to see them: almost mythical birds, circling high over a wood on the Fal Estuary. I watched them—naked eye, no bins, not back then—quite entranced. I had never seen anything so wonderful: and part of the wonder of seeing buzzards in those days was that, like all birds of prey, they were vanishing

fast. It seemed then that they already belonged to the past. We were rather apathetic about it: as if the loss of these birds was inevitable, nobody's fault, just the way of things, gone like Robin Hood and King Arthur, regrettable but unavoidable.

But now they're everywhere. In the West Country, buzzards are part of daily life for many people, and in recent years they have spread back across most of the rest of the country in robust numbers.

It goes back to pesticides. Organochlorine pesticides, most notably DDT, were in common and extensive use in the 1950s. These did more than kill insects. The stuff built up in the tissues of everything that ate the infected insects, and so the birds of prey that ate insect-eaters—i.e. all of them—were regularly taking heavy doses of the stuff. This, in turn, caused them to lay eggs with thin shells: eggs that were non-viable.

These pesticides were banned for agricultural use in the United States in 1972 and in Britain in 1986 and across the world in 2001. And since then birds of prey have come back in impressive numbers. The use, and eventual banning, of organochlorine pesticides demonstrates not only that unrestrained development can be damaging to the environment and to wildlife, but also that there's something we can do about it— and that in certain circumstances, we can win back something of what we have lost. That the environment is not a lost cause.

I was in Sri Lanka. I had turned thirty a few weeks earlier. I was travelling with a beautiful girl, now my wife. We were in the middle of a wide-open coastal wetland. This was not my natural habitat, not back then, for even if I had spent most of my childhood reading and dreaming about birds, I'd lost the faith in my teens. But on Cindy's insistence we were exploring the wilder places of this lovely island, and I had even hired a pair of binoculars to help us do so. There were some distant birds, so I raised the bins for a look.

And it was as if a nuclear bomb went off in my mind.

Rather to my surprise, I was able to identify the birds. Rather to my unutterable bloody amazement, they were avocets. In other words, they were birds of miracle, mystery and myth; birds that had haunted my boyhood but which I never expected to see for as long as I lived. I felt as if I was looking at a small herd of unicorns.

Avocets had become extinct as British breeding birds. But when the soft coast of Suffolk was allowed to flood during the Second World War to make any invasion from occupied Holland more difficult, this move of military good sense created the perfect habitat for avocets. And so, first at Havergate Island and then at Minsmere, avocets made the invasion that the Nazis never managed.

This recolonization was a miracle, or seemed like one. It was all the more vivid because it was a miracle of peace. The war was long gone. Now the country was going to be better than it had ever been before. It was going to be richer in wildlife, which meant that it was going to be a thousand times richer for us humans. The avocets' chosen places were soon owned and managed by the RSPB, who adopted the avocet as their

facing page:
AN AVOCET
(right) and a black-
winged stilt. Avocets can
now be found wherever
conditions are right
in East Anglia.

logo—and why not? This unmistakable bird might have been designed as a logo had it not already existed.

At first, access to these impossibly rare, impossibly valuable birds was strictly controlled. There is a crusty memo from the then warden of Minsmere, Bert Axell, to RSPB headquarters complaining that he had been forced to take more than thirty visitors to see the avocets in a single week. It was too much. How could we hope that the birds would continue breeding with such disturbance?

These days Minsmere is a visitor honeypot, with hundreds arriving every day. There is a vast car park with space for coaches, a tearoom and a shop. The BBC programme *Springwatch* is broadcast from Minsmere over three weeks. And the place is jumping with avocets. Avocets have spread out from their two original strongpoints and can be found wherever conditions are right in East Anglia. There is a wintering population of several hundred on the Exe Estuary in South Devon. It's possible to see avocets from the train between Exeter and Newton Abbot, so for enthusiasts that counts as a train tick.

The avocets' resurgence is a tribute to good luck, the resilience, resourcefulness and mobility of birds, and to excellent conservation work, not just in making places safe, but in managing and re-creating coastal marshland. Avocets show what can be done in conservation.

In the 1950s Eamonn Andrews presented the BBC children's programme *Playbox*. A regular part of the programme was the Six Clue Challenge, in which a criminal entraps himself under the interrogation

of Inspector Bruce by making six Very Foolish Mistakes. Can you spot them too? One of these playlets was called *The Case of the Painted Eggs*. It was about a dastardly egg collector who wanted to steal eggs from the ospreys nesting at Loch Garten in Scotland. The episode was written by my father and mother, quite close to the start of their careers in television.

Ospreys had become a national cause célèbre. They had been extinct as breeding birds in Britain since 1916. But in 1954 a pair turned up and nested in Scotland; in 1958 a pair moved in to Loch Garten. They were at once birds besieged: their nesting tree was surrounded with barbed wire and there was a twenty-four-hour watch on the site, known as Operation Osprey. People were appalled by the idea that there were individuals who—under the grip of collection mania—wanted to destroy the un-hatched chicks of a rare species.

Whose ospreys were they anyway? They belonged to themselves and they belonged to everybody. They certainly didn't belong to the criminal few.

The ospreys failed the first year but they came back. And since 1959 they have bred successfully almost every year. They were slow to recolon-ize, because of the problems with pesticides already touched on; there were still only fourteen breeding pairs by 1976, more than twenty years after their return. But things have moved on apace since then. There are now getting on for 300 breeding pairs in Britain. This has been helped greatly by a reintroduction programme at Rutland Water.

The RSPB own and manage Loch Garten and the forests around. The visitor centre there is another tourist honeypot; over fifty-odd years, more than two million people have dropped in to see the great birds. These days the ospreys can be observed not just through fixed telescopes but also by cameras and, inevitably, on the internet. Ospreys were wiped out in the nineteenth century, struggled back in the twentieth, and are a vibrant part of life in the twenty-first.

above:
THE OSPREY
was wiped out in Britain
in the nineteenth century,
but is a vibrant part of life
in the twenty-first.

Wildlife legislation is good news and bad news. It's a good thing that wildlife has legal protection, but it's a bad thing that such legislation is necessary. Once again it's the birds that lead the way, being the most obvious and the best-loved animals in the country.

The Wildlife and Countryside Act was a game-changer. It didn't solve every problem of interference and persecution, but it represented a serious change in the social and legal status of many wild species. It became law in 1981, and under its provisions it became illegal to kill or injure a wild bird or to damage a wild bird's nest. Certain species, like crows and wood pigeons, don't get the same protection because they are seen as agricultural pests. Other birds seen as quarry species are not protected during the shooting season.

The legal change has a great symbolic impact. But making a thing illegal is different from stopping it happening. And, obviously, the first problem is policing: you don't get too many squad cars on upland moors. If you fancy a pop at a bird of prey, the overwhelming likelihood is that you will get away with it. The second problem can be summed up in the old rhyme about the enclosure of common land that took place in Britain in the seventeenth century, under which public property became private:

> *The law locks up the man or woman*
> *Who steals the goose from off the common*
> *But turns the greater criminal loose*
> *Who steals the common from the goose.*

The problem is in the way we manage the land. In many places, agriculture is so intensive that birds can no longer make a living there

The problem is not just that people kill birds. The problem is also in the way we manage the land. In many places, agriculture is so intensive that birds can no longer make a living there. The small mixed farm is more or less a thing of the past; the mixture of pasture and arable in small, thickly hedged fields is a rare thing now. The more intensive the farming, the fewer birds we have.

Thus the good old farmer, archetype of British life and hero of a million children's stories, has become a villain: Farmer Birdkiller. It's a shocking development. Much of it comes down not to the tastes of individual farmers but to government policy, and the uncritical willingness to pursue a goal of agricultural efficiency at the expense of a countryside that's friendly to people and birds.

Recent figures show a 62 per cent decline in skylarks since 1970, and a 50 per cent decline in lapwings. Other species—tree sparrows, corn buntings, even starlings—have also declined steeply. It has come about because hedges have been ripped out, farmers have ploughed much closer to the field margins and insecticides are in wider use, so the birds have less to eat. Most crops are sown in winter these days, depriving birds of winter stubble, which was a valuable food resource.

In other words, there is a tendency to try to turn British farmland into Nebraska—but without vast acreages of national and state parks to compensate. It's a deeply damaging trend.

We live in a land that has far fewer skylarks in it. But there have, nonetheless, been some important shifts in the drive for efficiency above all else, and we will come to them in the last chapter. The apocalyptic vision of *Silent Spring* hasn't come to pass, in large part because *Silent Spring* gave us a warning that came in time. For once, Cassandra was listened to and believed, and her prophecies were acted on.

So spring isn't silent, thank God. Not entirely. But in a lot of places it's still pretty bloody quiet.

13

Birdwatchers, birders and twitchers

L eisure and loss. The combination of these two things changed the way humans look at birds, at least in the industrialized world. Both of these ideas would have been equally unthinkable—even incomprehensible—to our ancestors. Once humans gave up the hunter-gatherer life and took to agriculture, we became much better at surviving and expanding, in terms of both numbers and the areas of the world we were able to make a living in—but the price of these advances was to give up leisure for several thousand years. The price of human expansion was enslavement.

And as said before more than once in this book, the idea that nature was anything other than an infinite resource and an implacable foe was beyond the understanding of humans who struggled to grow crops and nurture animals in a hostile environment. All environments were hostile back then, even the gentlest and most fertile. Life was a struggle: that was accepted, and humans got on with it in ways beyond our imagining in these soft modern days in the developed world.

But when we went through revolutions in agriculture and industry the landscape changed, in both a literal and a metaphorical way. These dramatic changes were accompanied by a revolution in transport: railways were invented and roads grew both faster and safer. The towns were bigger, but it was now possible to leave them for a while. All of these things made it possible not just to visit the countryside but to enjoy the wild world. So people began to seek out the wild world for

previous page:
THE STONE CURLEW.
For caption, see page 277.

nothing but pleasure. And so we invented birdwatching.

People began to look at and for birds for no reason at all: not to read the future or fill the pot or to advance human knowledge. They looked at birds because they felt like it, because they liked doing so. It was a practice that answered a deep need. And perhaps the deepest need of all was for flight. Flight away from cities and crowds and industry and noise; flight in the lifting of the heart that comes with a nice sight of a nice bird; and flight as in the joy of seeing a living creature take wing.

Gilbert White, the Selborne parson (see page 211), is generally credited as the man who started it all off. His book, *The Natural History and Antiquities of Selborne*, was remarkable not just in itself, but also in its reception. It wasn't just a good book and the record of a good life of a great observer; it was also the book the world was waiting for. He caught the wave of longing for a wilder way of living that had been lost, for a reconnection with our birthright: the natural world. The book was a bit like one of the great David Attenborough television series: it brought the wild world into people's laps. Many people loved the idea of it as they read White. And a good many of them went out to seek the real thing.

The Romantic movement was another response to loss and it was, I suppose, another kind of birdwatching. The poetry of Wordsworth and Coleridge and, subsequently, of Shelley and Keats is filled with a conviction that really important truths can be found in the wild world, that human enrichment lay beyond the cities. It was the beginning of the realization that the breakneck pace of development was not an unambiguously good thing, and that without the wild world we are less than ourselves.

People began to look at birds—and listen to them, too, like Keats's nightingale—with no quantifiable personal advantage in sight. The birder's favourite is the Northamptonshire poet John Clare, two years older than Keats but published later. His work is full of genuine, accurate observation of the natural world. Keats's nightingale poem is about Keats; Clare's nightingale poem is about nightingales. Keats's nightingale sings in flight, which no living nightingale ever did; Clare's sings from cover, as a good nightingale always does.

The continuing popularity of such poetry reveals a social truth as well as a literary one. All at once birds were seen by many people as a source of pleasure. But how do you make sense of what you see?

The beginning—and for some people the end—of understanding birds is to identify the species to which each individual bird belongs. This is a process that makes you look and listen more closely, and pay better attention. By doing that, you do something to your brain. You tune in. You set yourself up to receive and respond to the stimuli you have opened yourself up to. What was that? It's the first question you ask about a bird. It's the question that turns you into a birder.

Perhaps this is the moment to consider terms. Birdwatcher? Birdspotter? Birder? Twitcher? The last is a specialist term, which we'll come to later in this chapter. For the rest, as far as I am concerned, anyone who has ever looked at a bird and wondered what species it was is a birder, and a watcher and a spotter as well, if you like.

Anyone who has ever had a conversation with a birder in a pub garden knows the way that, while paying attention to the conversation and indeed contributing to it (if not actually monopolizing it), the birder's eyes will constantly flick to the sky, and his ears will also provide all kinds of important information. Here's Matthew Engel, recording such a conversation in his excellent *Engel's England*: "'The whole point," he was saying, "is that a reed bed wants to become an oak forest. *Shoveler*. Left to themselves the reeds would slowly deposit humus and dry up and eventually the scrubby stuff would *reed warbler, no, sedge warbler* take over…"' That was me he was quoting.

below:
BEWICK'S SWAN
in an engraving by
Thomas Bewick's son
Robert Eliot Bewick.

So how do you start? The British people started with *A History of British Birds* by Thomas Bewick, the first volume of which was published in 1797. He illustrated this work with his brilliant wood engravings, in which he used a radical and advanced technique that gave real detail to the images. Bewick is commemorated in Bewick's swan, a fine bird (see page 103) that comes down from the Arctic to spend its winters on British shores.

His work was followed by that of George Montagu, army officer and naturalist familiar to birders because of Montagu's harrier, Britain's rarest breeding bird of prey, and arguably the most elegant. Montagu produced the *Ornithological Dictionary*, published in 1802 and revised in 1831. In this he accurately catalogued all the species you are likely to find in Britain, and describes them in meticulous, feather-counting detail.

Bewick and Montagu between them did a fine job for scientists and perhaps an even finer job for the laity: for people who found increasing pleasure and increasing meaning from increasing knowledge and understanding of what they were looking at. Here was information: good, hard, accessible. Here was the wild world. Let the joy begin.

right:
MONTAGU'S
HARRIER.
Britain's rarest—and arguably most elegant—breeding bird of prey.

It was to be a couple of centuries before the term was invented, but it's also essential to note that these primordial works of Bewick and Montagu were a celebration of biodiversity. Here was a challenge both to the observer's skill and the reader's mind. So many different species! Not just nameless little brown birds and a generic bunch of birds singing away in cover. Now it was clear to all that each species had an identity and a meaning that was separate from that of every other species. This, in short, was how the world works. No wonder people were fascinated.

Gilbert White showed us that the willow wren was in fact three different species (see page 211). His *Natural History and Antiquities of Selborne*, along with Bewick's and Montagu's books, told us that the wild world is more complex and perhaps more beautiful for that reason: more complex and more beautiful than we had ever imagined.

But birds represent a kind of diversity that is within our grasp. If you look for micromoths, there are more than 1,000 to choose from in this country alone, and they're fiendishly difficult to tell apart. Birds come in a level of diversity that suits the human mind, in numbers that we can deal with intuitively, numbers that make sense. On a good half-day's expedition, you might encounter fifty species if you were lucky, and that's a number within most people's grasp. On a mad, intense day at the right time of year and in the right places you might just get the century up: a number that's part of the way we think.

This revelation of diversity opened the way to a strange human impulse: the desire to collect. Collection is a mania that strikes people in all sorts of circumstances and it is inspired by many different things. The glorious plot of *The Code of the Woosters*, by P. G. Wodehouse, centres on the collection mania for Georgian silver. At the start of the book, Bertie is given the task of entering a shop and examining an eighteenth-century cream jug in the shape of a cow. He is instructed to scoff 'Modern Dutch!' before walking off in a marked fashion.

This same sort of collection mania has always been inextricable from the human fascination with birds. In Victorian times, a collection of dead birds was very much the thing, and collectors vied with one another for the extent and the beauty of their collections and the rarity of their most prized specimens, and as we've seen, that was the last straw for the great auk (see page 228). A stuffed bird—a collection of one—was a standard aspect of interior decoration, spreading into the lower middle classes. Leopold Bloom, hero of James Joyce's *Ulysses*, has a stuffed owl in his sitting room, a wedding gift, much prized.

The collecting of eggs was a traditional country pursuit for boys across generations. It combined the skills of the hunter and the desire for completion: to have an egg from every British bird or, failing that, at least a few more than your nearest rival. Bill Oddie, as sound a conservationist as has ever been hatched, freely admits that his love for birds began with egg-collecting. He gave it up while still a boy 'and instantly became a better person'.

The practice of egg-collecting continues as a secret cult, shabby and furtive, its illegal nature doubtless adding to the thrill. As a result of this,

Leopold Bloom, hero of James Joyce's Ulysses, *has a stuffed owl in his sitting room, a wedding gift, much prized*

when the chough came back to Cornwall and when the cranes came back to the Fens, there was need for a twenty-four-hour watch on the nest sites. The eggs would have had incomparable value for such people, like the gloaters who turn up in thrillers with secret collections of stolen paintings.

Are you a twitcher, then? Most birders get sick of explaining that they're nothing of the kind. A twitcher is a birder—often a very good one—who is in thrall to the love of the list. Their British list is usually the most important thing in their lives, and they will travel immense distances at great inconvenience to see some poor bird that's been blown in from Siberia or from the far side of the Atlantic.

It's important to understand that a rare bird, in such terms, is not necessarily one near extinction. It's far more likely to be a bird that's common in one place, but has somehow got itself transferred to another place, where it is not common at all. Mark Cocker, in his enjoyable *Birders: Memoirs of a Tribe*, talks of a pair of twitchers who learned that a nighthawk—an astonishingly rare bird in Britain—had turned up on the Scillies. They were in the United States at the time and rescheduled their flights to get home, went straight to the Scillies and eventually found it. A nighthawk is an American species related to nightjars; you can find them in the United States. And the twitchers heard about this amazing bird while they were on Cape May and surrounded by a flock of forty-nine nighthawks. Why do such a thing? It looks like pure madness, but it's nothing of the kind. It's sport, it's competition. They wanted the bird on their treasured British lists, and they didn't want other twitchers to have the birds while they didn't. Hence the trip, and hence the feeling that it was bloody well worth it too. Me, I'd have stayed on Cape May—but it wouldn't do if we were all made the same.

This strange cult has received much media coverage; we British always rather relish evidence of our own barminess. Twitching is now accepted as a traditionally ludicrous aspect of national life, like Morris dancing and bog-snorkelling. And it's all fair enough: any way you enjoy your birds without harming them is good by me. But the idea that twitching and birdwatching are one and the same thing has put many people off the joys of birding.

I wrote a book called *How to be a Bad Birdwatcher* in direct response to this misapprehension. It was an attempt to legitimize the pleasures of looking out of a window and seeing a bird and feeling a little better about your day and your life by doing so. Birding is not necessarily about obsession, and expertise is not strictly essential. The process of acquiring greater knowledge is rewarding, sure, but an innocent delight in the bird in front of you is also something that matters.

There are suggestions that twitching is becoming slightly old hat these days: that great rarities no longer attract the vast crowds of a decade back. Perhaps one reason for this is that digital photography has provided a new way of expressing collection mania. Bird photography once required hours in a hide with enormous, weighty and highly expensive pieces of specialist equipment, but these days anyone can have a bash at it. The gear is light and comparatively inexpensive, and the material you gather is more or less free and entirely under your control. You can use a digital camera in partnership with a telescope—a process called digi-scoping—and reach out to take a picture of a small bird from an

facing page:
THE NIGHTJAR
is a crepuscular bird,
searching for insects at
dawn and dusk.

immense distance and fill the frame while you're doing so. You can then examine the bird at your leisure with all kinds of reference material in book form and online.

Old-school birdwatchers are inclined to decry this. Taking down an exhaustive field description when you come upon an unusual bird has always been one of the great traditions and great skills of top-level bird-watching. But you no longer have to do this. You just get your picture and send it to your county or your national rarities committee and the job's done.

For a twitcher it's enough to see a bird; the collection is entirely intangible. But the growing numbers of photographers want to capture—please note the word—an image as well. It's a hunting thing, then. Ever since birdwatching became an acceptable pursuit, a good deal of it has been driven by the urge to collect.

Binoculars gave all birdwatchers something that only exceptional observers like Gilbert White had before: a genuine intimacy with the subject

You might think that birdwatchers are boring when they talk about birds, but that's nothing. You ought to hear them—us—talking about optical glass.

But then binoculars really are the most astonishing piece of kit. Binoculars remove the wall at the back of the wardrobe and take you through to the wonderful country of the birds. They don't bring the birds closer to you—they bring you closer to the birds, taking you into the tree, across the sea to the ledges of the cliff, out into the middle of the lake, and above all, they take you into the sky to fly with the birds. Every pair of binoculars that any birdwatcher ever looked through was not really manufactured by crack German precision engineers. They were actually made in Diagon Alley, by Ollivanders, the shop that sold Harry Potter his wand.

Binoculars change the world. They change your understanding of what's possible. Binoculars gave birdwatchers something that only exceptional observers like Gilbert White had before: a genuine intimacy with the subject. As a bonus—and a not inconsiderable one—binoculars meant that birds could now be enjoyed and technicalities of species could be addressed without any need to kill them. Binoculars allowed the first birdwatchers to revel in wildlife without needing to make it wilddead.

Binoculars began as opera glasses. These were developed in the eighteenth and nineteenth century so that opera-goers could see the stage and the singers better. My guess, from reading Proust, is that their more significant use was to allow opera-goers to watch each other during the intervals, and to look for important figures in society. *Mon dieu*—is that really the Duchesse de Guermantes? *Mais bien sûr*—I'd know Oriane anywhere. Look at her feathers...

These toys were then adapted for outside use and, like so many advances in technology, this one came about because of war. Binoculars—actually called 'field glasses'—were issued to British troops during the Crimean War of 1853–6. They weren't much cop. That was because if you want proper magnification you need the light to travel for some considerable distance between the lenses, and there's a limit as to how long a portable bit of kit can be. But in 1854, Ignazio Porro, an Italian inventor, patented prismatic binoculars, and that changed everything. In this system—used ever since—you get the length of light you need by bending it around a series of prisms.

Stephen Moss tells the story in his excellent *A Bird in the Bush: A Social History of Birdwatching*, explaining that the Zeiss Feldstecher went on sale in Britain in 1896, costing up to £8: serious money in those days. The subsequent story is of ever-better instruments, and also of ever-cheaper instruments. These days you can buy a pair very cheaply indeed, or you can spend a fortune and buy the best. Either way, binoculars are available to transport you across space and bring you closer to the meaning of birds.

The moral, then, is to get the best pair of bins you can afford and never leave them at home

In the 1980s it was generally acknowledged that the best binoculars ever made were Leitz Trinovids. I was an anti-snob at the time and scoffed at the idea of paying silly money when there was something just as good for an eighth of the price. But one day I was in an optical shop and I thought I'd try a pair of Trinovids just for the hell of it. Mostly so I could sneer and say, 'See? What did I tell you? Not worth the money, right?' I stepped into the street and raised them to the building opposite for a test drive: and damn nearly dropped them I was so amazed. And then bought them, of course. You get what you pay for.

'I am observing sroo my powerful Gestapo binoculars,' says Herr Flick in the great sitcom *'Allo 'Allo*. Binoculars are always supposed to be 'powerful'. But power is actually the last thing you need.

'How many magnifications?' people ask, when they feel the need to ask a polite question about your bins. But magnification is not actually the point. You don't want huge magnification, because that would only exaggerate your own involuntary movements. You couldn't hold a pair with twenty mags steady enough for long periods to make them a useful tool. What a great pair of bins will give you is clarity, brightness and a three-dimensional image.

When someone looks through a pair of bins in a film you always see his point of view in a figure-of-eight shape. But a good pair of bins gives you nothing of the kind. You get a circular image. The image from one side—one eye—is superimposed on the image from the other, and with immense precision, and that's what gives you an image in three dimensions. I am observing sroo my *precise* Gestapo binoculars. The light-gathering qualities of a good pair of bins are so good that in low light you can see better with the glasses than you can without them.

The moral, then, is to get the best pair of bins you can afford and never leave them at home. But please do your best not to talk about them.

Birding is a shared passion and a shared joy, but you wouldn't think that from listening to some birders. They always seem to be falling out with each other. And one of the eternal divisions between birding people is the science. Just how scientific are you? There are those who keep beautiful notebooks full of accurate observations, written up after every birding expedition. Some take detailed notes of behaviour; most make a point of noting weather, wind direction, temperature, season. But there are plenty who just go out and see a bird and say, 'Fab. Nice bird, eh?'

Some people will see a buzzard and work out that it's actually a rough-legged buzzard. They will then take the notes, take a photograph and submit the sighting to the county bird recorder. Others will say, 'Yeah, I reckon that's a rough-legger all right. Nice!' And leave it at that.

Meanwhile, plenty more will admire a big bird of prey and explain when they get back that it made their day and it was probably a buzzard, and others will say it was probably some kind of eagle, and it was absolutely brilliant.

I have no quarrel with any of the above approaches. The data collected by birders can be used for scientific purposes. The excellent organization the British Trust for Ornithology (BTO) does a great deal to collate such data, so as to tell us often surprising and sometimes deeply disquieting things about the numbers of birds we have in this country.

And that's not only great, that's essential, and all the people who send in good information to the BTO are to be respected and thanked. But that doesn't mean that a bad birdwatcher can't have a great day out and tell nobody at all about it apart from a few friends.

Birdwatching carries on as a great democracy. Birds are for all and can be enjoyed by all, in many different ways. Note that word: enjoy. Birdwatching is something many people do because they enjoy it; it enriches their lives. It can be the guiding passion of your life, or it can be something you do on a single day of your annual holiday. Either way it's legitimate and it's good.

In 2015 I spent a day with a professional conservationist—Nigel Cowan, project manager of the East England Stone Curlew Recovery Project—seeking out stone curlews in East Anglia. That same year I took a boat trip with children, babes in arms, dogs and grannies to see the gannets at Bass Rock, and I was almost the only person among the fifty or so on the boat who had binoculars. And we all came back filled with joy. Both days were absolutely great.

left:
THE STONE CURLEW is a rare summer visitor to Salisbury Plain and the Brecks of Norfolk. Stone curlews are night foragers: hence the mad eyes, goggly for improved night vision.

I was, I think, five or six when my father gave me a copy of *The Observer's Book of Birds*. It cost him five shillings and I read it to bits, until I had to buy myself a new one. The book was first published in 1937—initially at half a crown; it sold more than three million copies and was the standard bird reference book for about fifty years.

It was tiny, fitted comfortably in the hand and had illustrations, one per page (a good many of them actually in colour), of all the common British birds. It was a reasonable price, a reasonable size, reasonably informative and it offered the reader a reasonable level of diversity. It launched a million birdwatchers, maybe many more.

More than anything else, it democratized birding. Most people had leisure time, anybody could afford half a crown, most people had access to open spaces even it was only the garden, the park or the common. We could all feed the ducks and now we could know the difference between the ducks with the green heads and the black-and-white ducks that always seemed to swim further out.

I then learned my multiplication tables. This is the only feat of mathematics I have ever mastered, and I can still tell you what seven eights are to this day. I did so because my parents had promised me a present if I ever pulled it off. I got *A Field Guide to the Birds of Britain and Europe*, put together and illustrated by Roger Tory Peterson.

So I've always had access to reliable information about birds, and it has been added to and improved over the years. I've just done another quick count of what I have on my bookshelves, because I've had a bit of a reorganization since the last mention (see page 92), and there are now sixteen field guides to British and European birds on my shelves,

The Observer's Book of Birds *sold more than three million copies and was the standard bird reference book for about fifty years*

and thirty-seven for the rest of the world. There still don't seem to be as many as I remember... A few of them must still be lying about somewhere else. Most of these books are marvellous. I have apps on my phone that not only provide all kinds of information and illustrations relating to birds in Europe, and for that matter in southern Africa, but also give you a recording of their calls. The twenty-first-century birder has every possible chance of identifying any bird in the world.

Information, equipment, leisure, travel: it's all there, better than ever before. More and more birders. Fewer and fewer birds.

14

*Feed the birds,
tuppence
a bag*

B reakfast in the garden is a treat for any English soul, especially in March. There was a little notice on each table that asked customers not to feed the birds, which naturally meant that customers always *did* feed the birds and the birds knew it. So I took that as an invitation, sprinkled a spoonful of sugar across the tablecloth and waited for both the waiter and the waiting birds. As can be expected, the birds arrived first: charming tit-sized things called bananaquits, with yellow bellies and heads like little badgers. I was in Barbados, covering the cricket in the days when I was a sports writer for *The Times*, and this seemed to me the best way to start the day. The bananaquits did all right in the course of that Test match; better than the England team, certainly.

Feeding these nice little birds, having these nice little birds on the table, was a small but real joy and I sought it without hesitation. I'm convinced that feeding the birds is a basic human urge and that it goes back to the dawn of humankind. There is a pleasure in giving, sure, but there is also a pleasure in taming: in conjuring these beings down from the sky and having them so close. This crumbling of the wild/tame barrier is a meaningful thing; it's a confusion of the boundaries that lie between human and non-human life. This is a fair trade. The human emphatically gets something in return for their generous act of donating food.

Our earliest ancestors would certainly have been surrounded by opportunistic birds when they ate, waiting for what they left. But— and it's almost a definition of their humanity—one or two among the

previous page:
MALLARD.
For caption, see page 289.

gathering would surely have thrown them a morsel and enjoyed the sight of a bird dropping out of the air to seize it and then make his escape. So next time you might throw the morsel just a little closer...

This would have been the way that birds were originally domesticated; pigeons eventually setting up home within easy distance of the larder, and red junglefowl eventually accepting complete human dominance in exchange for a new set of certainties.

But many birds that benefited from human feeding were not domesticated, and perhaps they weren't even caught and eaten so much. Birds— some species, anyway; mostly the omnivorous and opportunistic kind— would always have been attracted to human settlements, and would have become both a nuisance and a pleasure: sparrows nesting in the thatch, crows gathering round the midden, pigeons roosting in the surrounding trees. And it was only natural to fling them a dole and enjoy their proximity as they dropped down. Birds can be more confiding than most other animals, because they can fly away in an instant. They can be closer to us than other wild things because they can escape so easily.

People with the right kind of attitude can easily persuade wild birds to take food from their hands. It just takes a little time, a certain aptitude for stillness and, perhaps more than that, good vibes. I remember a walk at Gilfach, a Welsh nature reserve in which the warden, Pip Amos, had persuaded the blue tits and great tits from around his home in the wood to take food from his hand. As a result, he was fair game anywhere within a quarter mile of the cottage. As we walked he was routinely buzzed by eager little birds. Had I been alone they would have kept clear, as they did when I went out by myself later in the day. But since I was with him

I was trusted—trust by association—and they took mealworms from my fingers too, perching on my hand to do so and digging in their sharp little claws without compromise.

The feeding of birds has become part of twenty-first-century life. In some countries in the developed world, most notably Britain and the United States, feeding birds in the garden is normal, expected behaviour. In the suburbs a person who doesn't put out food for wild birds is more likely to be regarded as the eccentric than the person who does.

And it's become an industry. In the winter of 1890, British newspapers asked people to put out food for the birds: bread, kitchen scraps, that sort of thing. In 1910 *Punch* magazine said that feeding birds was 'a national pastime'. And since then, feeding the birds has become nothing less than an expression of who we are and what we stand for.

Feeding the birds represents an acceptance that we humans need to put ourselves out a little if we want non-human creatures in our lives. It shows that we accept that birds depend on human goodwill in almost every aspect of their lives. It also represents another truth: that we humans find the proximity of wild birds a deep pleasure. It has come to feel like an evolutionary strategy: birds exploit human love in order to get food, and the food they get is not begrudged. It's a fair exchange based on solid principles. It might look like a precarious strategy to build a life on, but it's robust enough to have worked for more than a century.

Inevitably, some people have claimed that the supplementary feeding of wild birds is in fact a bad thing. It has been suggested, for example, that it fosters dependency; that garden birds get used to human goodwill and then they can't do without it. But that's true, to a greater or lesser extent, for every bird on the planet.

There have also been claims that feeding affects the natural patterns of distribution: but then, what's natural? Do we count humans out of the natural processes? If we do, then it's a hard life for the birds, because humans affect practically everything that happens on this planet, including the weather.

It's also been claimed that feeding affects the density of population, disrupts migration patterns, promotes malnutrition through incorrect feeding, spreads disease and makes the birds vulnerable to domestic cats and window strikes. And while there is truth in all of this, it's generally reckoned that the good by far outweighs the bad. And it's also true that birds that can adapt to humanity have a far more secure future than those that can't.

In Britain the main bird charities are all in favour of feeding the birds. The RSPB and the British Trust for Ornithology both recommend it and go to some lengths to tell you the best way to do it and the best food to offer. Bird-feeding is an activity that offers specialist equipment, in terms of feeders, and a remarkable range of feeds.

CJ Wildlife offer, among many other treats, a specialist robin blend, muesli mixed with mealworms, pinhead oats, suet puddings in the shape

of a Christmas tree or a (pink) heart, live giant earthworms and peanut cake. They will also sell you every kind of feeder you could imagine— hanging, free-standing, squirrel-proof—and even a specialist device for dispensing nyjer seeds (a favourite with finches).

And if that seems mildly amusing, it's also a serious business. It's been claimed that half the adults in Britain feed the birds. In the United States it's been calculated that more than 55 million people feed the birds and they spend three billion US dollars every year in order to do so.

Towards the end of the day at REGUA it was my custom to take a glass of beer at sundown; not that this makes REGUA unique—at least, not in that way. This is the Reserva Ecológica de Guapiaçu in Brazil, a place mentioned earlier (see page 95), which looks after and restores the Atlantic Forest, the most trashed habitat on earth. I was there because of my association with the World Land Trust, a partner organization based in Britain.

I would naturally take the beer in the same place overlooking the bird feeders, and every evening, as I took my first drink of the day, a bird would come and take his last. He would always arrive with a great roar of his wings, a wonderfully melodramatic arrival calculated to terrify any other bird within earshot. Then he would sip at his ease from the bottles filled with sugar water, poised before the openings as if the air was a solid thing you could perch on. This was a swallow-tailed hummingbird, and he was accustomed to making these regular performances throughout the day.

It was a great moment for humankind when someone discovered that

facing page:
ROBIN REDBREAST.
In 2015 it was voted
Britain's national bird.

hummingbirds would take artificial nectar from artificial flowers, allowing these gloriously improbable birds to come right down and breathe the air with us. I'd try the same thing at my garden in Norfolk, but I fear that my neighbours have been scattering hummingbird repellent.

The RSPB gets a few calls every summer from people who have been thrilled to find a hummingbird in their garden feeding from the flowers, drinking nectar and hanging in the air as if weightless. But it's always a hummingbird hawk moth: an insect that has evolved the same technique as a hummingbird by a completely different evolutionary route. It's another example of convergence (see page 228), and it's the evolutionary equivalent of great minds thinking alike.

Feeding the ducks is one of the rites of childhood. It was estimated that in 2014, English and Welsh people fed six million loaves of bread to the ducks. Certainly, feeding the ducks is one of the small pleasures of being alive. But, alas, though it gives pleasure to both parties, this is one time when the begrudgers have a point. Feeding bread to ducks is in fact a bad thing—especially when it's white bread. It fills them up without giving them proper nutrition, and what is left uneaten, in its small way, pollutes the waterways. By all means feed the ducks, but give them oats, corn and peas. Allow a packet of peas from the freezer to warm up by the time you reach the river or the lake in the park, and then you can enjoy your generosity and the birds' proximity with a clean conscience.

My old friend Tone Horwood was brought up at Churchill Gardens in Pimlico, London, so naturally visiting St James's Park was a regular childhood treat. In one of his earliest memories Clementine Churchill, wife of Sir Winston, addressed him with great kindness. 'Are you feeding the dilly-ducks?' she asked. Tone answered in the insufferable manner that he would certainly use to this day: 'They're not dilly-ducks. They're ducks.'

facing page:
VULTURES
ride the air with
scarcely a flap, cruising
so close to the sheer
walls of the mountains
that you'd think
they'd brush the wall
with their wings.

I was making a birding trip to Wales one winter, so of course we paid a visit to Tregaron. There was something eerie and otherworldly about it. It didn't feel quite like the normal Welsh countryside. It was as if we had gone through a door into a place that looked the same but felt quite different. The reason was that there were large birds perched in the tops of the trees. It was like looking for stars at dusk: the more you looked, the more you saw. And it was quite clear that these were red kites. These birds had become extinct as English breeding birds during the great Victorian persecution, but they hung on in Wales. And with the changes in the laws on pesticides (see page 252), they have made a comeback. All the same, you don't expect to see them gather in crowds.

There was an explanation for this. As the light began to fade in the short winter day the birds suddenly became restless. This restlessness was associated with the arrival of a woman, middle-aged and wearing a sensible skirt—the sort you'd expect to find running a stall for the upkeep of the local church. What she did was to scatter a good deal of rather nasty meat about the place: butcher's leavings.

And down they came, about fifty of them: broad red wings, sharply forked tails, languid flight and ravenous appetites. They ate their fill from this bloody and benighted bird table—well, feeding station, since no actual table was involved—and then returned heavily, with full crops, to the trees for a long, chilly night of delicious digestion. The lady with the meat then passed among the small group of spectators and collected some money from us all. It was for the upkeep of the local church.

The custom continues to this day. During the winter months the kites are fed every day at two o'clock, and you can not only witness the spectacle, as I did, but also learn more about the red kite, and other local wildlife, by paying a visit to the Tregaron Kite Centre and Museum.

'They know the van,' I was told. 'And when they see it, they follow it.' The van comes from the local slaughterhouse and makes the short journey through the Pyrenees to the appointed spot and dumps bodies unwanted by the meat trade. Then the vultures come swooping down, and once on the ground they feast and squabble and hiss at each other. It looks like a convention of jabberwocks.

The vulture population of Spain—mostly griffon vultures and the much smaller Egyptian vultures —crashed after there was an EU ruling about dead cattle. It occurred during the outbreak of BSE (bovine spongiform encephalopathy, nicknamed mad cow

disease) in the 1990s. Under the new ruling, cattle that died had to be cleared up; it had formerly been the custom to allow free-ranging cattle to stay where they fell and to become part of the economy of the soil.

So here was a compromise: and as a result there are, once again, plenty of vultures to be seen in the Pyrenees. They ride the air with scarcely a flap, cruising so close to the sheer walls of the mountains that you'd think they would brush the wall with their wings and tumble out of the sky. But of course they never do. It's as wild a sight as you can find in Europe, and you have to remember that it happens because of the way we humans have arranged things. The vultures exist in their present numbers out of human favour; out of human desire for the wild.

Vultures exist in their present numbers out of human favour; out of human desire for the wild

The California condor—as impressive a bird as you can find with a wingspan of 3 metres (getting on for 10 ft)—went extinct in the wild in 1987. The reasons were habitat destruction and poisoning from the lead they ingested when they ate animals that had been shot. Wild-caught individuals were taken to zoos and a captive breeding population was established. Eventually, California condors were released into the wild in half a dozen locations in their ancestral range of California and Arizona. The process of re-establishing the birds involved artificial feeding: putting out dead animals for the birds to feast on.

The Wildfowl and Wetlands Trust's first reserve was at Slimbridge in Gloucestershire, on the Severn Estuary. I was once lucky enough with the timing to arrive with the Bewick's swans (see page 103). This was a glorious sight, the lake crowded with these huge birds. One of the reasons they are so faithful to the site is because every evening they put out large quantities of corn for the birds to eat.

The RSPB doesn't feed swans on its own reserves, having a different policy about when it's appropriate to intervene—though many RSPB reserves have bird feeders for the little birds and, as we've seen, they are very keen for you to feed the birds in your own garden.

When is feeding interfering with nature? Where do you draw the line? It is slightly less satisfying to know that the bird you are looking at is not entirely wild, in that it accepts food from humans. But if we don't feed them, we will have fewer birds to enjoy.

Once you start feeding birds in your back garden, you never know where it will end. Before long you'll be giving up the whole garden to birds… and that's exactly what people are doing. After all, why stop at putting out a few peanuts? People are now managing entire gardens for the benefit of birds and other wildlife. When people decide to put in a shrub, many are likely to look for one that will bring birds to their garden, planting, say, a cotoneaster for the berries, which look very jolly in the colder months and will also attract thrushes, including the fieldfares that come to this country from Scandinavia to spend their winter with us. And you could get lucky and have a party of waxwings dropping in to strip the bushes: as thrillingly exotic a bird as Britain can show you.

As more and more of the country gets concreted over or rendered useless for wildlife by intensive farming, so people are beginning to use their gardens as ad hoc nature reserves. Naturally this works best for birds, who can fly from garden to garden, though there are many examples of people leaving gaps under gates and holes in fences to allow hedgehogs to have easier travelling on their nightly forays.

It's been claimed that the largest nature reserve in the country is the back gardens of Britain. Certainly it's true that the more people who manage their gardens in this fashion, the more effective they will be as a wildlife reserve, and the richer our land will be in wildlife.

The garden pond is now a standard feature of many a garden: the centrepiece of a place that is designed for the pleasure of humans and the maintenance of wildlife. To see a blackbird taking a bath in a pond that wasn't there before is a rather wonderful thing. Gardens have become a crucial human–avian interface; places where humans can easily and without fuss take a daily pleasure in the wild world.

Many people have come to support conservation organizations even though they would never think about visiting a nature reserve, still less calling themselves naturalists or birdwatchers. They just like looking out of their windows and seeing wild birds—and who can blame them?

Very few members of the RSPB consider themselves birdwatchers, and yet they support the organization and are part of a million voices that speak up for the conservation of wildlife through the organization. They have come to appreciate the importance of wildlife by means of the bird table. The bird table is not only an important resource for wild birds;

It's been claimed that the largest nature reserve in the country is the back gardens of Britain

it also brings people to conservation in vast numbers, and their voices and their financial support are immensely helpful in the daily battle to persuade people in positions of power to take wildlife conservation seriously. A million voices have to be acknowledged.

15

*Hope is
the thing with
feathers*

Are the birds of Britain and the world half destroyed? Or are they half saved?

Yes. Both.

So perhaps it comes down to what kind of person you are. I could fill this final chapter with a thousand horror stories of indigence, casual slaughter, purposes mistook, criminal carelessness, deliberate destruction and devastation. But I could just as easily fill it with a thousand more stories about great people doing great things in great places: fostering wild populations, restoring lost habitats, filling the air with birds.

The state of the wild world invites the wildest of mood swings. It seems that hope and despair are forever racing neck and neck. In Borneo I travelled in an hour from areas where the forest had been utterly destroyed to make way for palm-oil plantations, to areas of glorious rainforest full of orang-utans, the charming pygmy elephants of Borneo and the impossible hornbills. And after that I visited an area of land that had just been acquired by the great local conservation organization Hutan (see page 307), and saw how it joined together two essential pieces of the damaged forest. Here was a healing of harms.

So am I an optimist or a pessimist when it comes to the birds and their continuing relationship with humans? I can give you an unequivocal answer on that one.

Yes. Both.

previous page:
RHINOCEROS
HORNBILL.
For caption, see page 306.

'"Hope" is the thing with feathers.' Emily Dickinson's words prompted Woody Allen to entitle a collection of his writing *Without Feathers*. This book has plenty of feathers—though not as many as I would like.

Here's how you do an ornithological survey. You walk along a set route with a clipboard in your hand and you make a note of every bird you come across. You walk the same route more than once, in an attempt to weed out errors. I once recorded six turtle doves on a visit to a nicely managed farm a good few years ago, and always wondered if there was actually only one, but very mobile. A subsequent visit turned out to be impossible, so I'll never know. Accurate surveying requires repetition.

We watch birds for pleasure because they are the easiest living things to come to our attention. That also makes them especially relevant to science and conservation and the gathering of information. If you want to work out how good a place is at sustaining life, the easiest thing to do is count the birds. They're easier than any other taxonomic group: easier to find and easier to identify. And the numbers you find are a clear indication of the nature of the place you survey.

You do most of this work by ear. That's especially true in spring, when you're surveying breeding birds, and as I hope you are now aware, many breeding birds sing, and the song announces their identity as members of a species and probably as individuals. It doesn't matter if you can't see it; if it sings like a willow warbler then you know damn well it's a willow warbler, so you mark it on your map.

I have a permanent relationship of acute guilt with that excellent organization, the British Trust for Ornithology. They collate all the information from volunteers and produce accurate figures about the bird population of the country. I have done some odds and ends for them, but I have promised to take on an area a few miles from my place in Norfolk. So far, I have failed to make a single survey. Next spring for certain. As I promise every year.

I have made a great advance in the past twelve months, though. I have learned who owns the place, and I have even acquired permission to make the survey. After all, I need to use such modest skills as I have in the right way. Wish me luck.

The survey area is on private land, but it is beautifully managed for wildlife. If I ever manage to complete a survey, I will be able to report good news back to the BTO headquarters on the other side of the county. A pity everywhere isn't like that.

If it sings like a willow warbler then you know damn well it's a willow warbler, so you mark it on your map

Co-operation is a fine thing.

In 2013, twenty-five organizations devoted to conservation and the recording of information about the wild world got together to produce the State of Nature report. It is a staggering document. It should have stopped the country in its tracks. It showed beyond all doubt that, whatever politicians and governments and statutory bodies want to say, we are losing the wild world at a terrifying rate, and if we don't accept the fact and do something about it, we'll soon have nothing left but rats and feral pigeons.

This was laughed off as just another scare story: the sort of thing that conservationists with bees in their bonnets come up with every so often.

But it was nothing of the kind. The State of Nature report is an unprecedented gathering together of meticulously recorded and detailed information covering 3,418 species of living things. The research had been done on vascular plants, bryophytes, fungi, butterflies, dragonflies, other less obvious insects, other invertebrates, marine animals and all vertebrates. And the most obvious of them all were of course the birds.

The figures showed horrifying changes that have taken place over the last fifty years. It turned out that 60 per cent of all species across the entire range of creation had declined. Of these, 31 per cent had declined sharply. It is also clear—the figures don't exist but the anecdotal evidence is utterly compelling—that the fifty years before the survey were also a time of decline. We have, quite literally, lost an immeasurable amount of life from our country.

There was also an indicator calculated from a list of 155 species that have been closely observed and documented, and many of these were of course birds. Of these, 77 per cent had declined over the last forty years, with little sign of recovery.

Another 6,000 species were assessed using what's called Red List criteria, following standard procedures laid down by the International Union for Conservation of Nature. This showed that one-tenth of all these species are at risk of becoming extinct in Britain. It was also shown that there were many significant alterations in abundance and distribution, reflecting the changes in the climate.

The main reason for these declines is simple enough to work out: the destruction and degradation of natural habitats. We're killing the wild places, and by doing so we're killing the wild world.

It has long seemed to me that there are two kinds of problems in conservation. There are problems that make you sad, and there are problems that make you angry.

The sad kind are the most important. Many of these problems have their roots in the enormous issue of human overpopulation. Thus we have problems with spreading towns, demands for a better life, problems of industry, problems that come from intensive agriculture and the perpetual question of what our countryside is for: whether it should be nothing more than an outdoor food factory, or whether it has a role in human happiness as well as human sustenance.

The problems that make you angry aren't complicated at all. They come from wanton destruction by a few recklessly selfish individuals.

The war with such people shouldn't dominate the conservation agenda. We have more and greater things to do. Better water management is a bigger issue than most, but it's a rather unglamorous and complex one. These important problems occupy many conservationists who do their bit for the wild world in windowless committee rooms and hotel lobbies.

All the same, there's no need to let the bastards get away with it.

There are problems that make you sad, and there are problems that make you angry

In 2001 there was an outbreak of foot and mouth disease in Britain. As a result much of the countryside was closed: no walkers, no wanderers, nothing. On the upland moors where the sheep graze in impossible thousands, there is a strange industry, already encountered in these pages. It's about big money. People will pay dizzying sums to kill grouse:

£4,500 for a day's shoot for one person is not unusual. The more birds they can kill in a day, the happier they are.

It follows, then, that the more grouse there are on the hills, the more money there is for the people with the shooting rights. It follows, then, that the people who run grouse moors don't like birds of prey. In particular, they don't like hen harriers. And in those glorious days of foot and mouth, when the cat was away and there were no witnesses for miles, the hen harrier population of the country mysteriously plummeted.

The killing of birds of prey is illegal, but policing vast acres of uplands is not possible. The decline of hen harriers has changed the nature of the conservation movement. It has become a lot less polite. It has become angry. This anger has spread to the wider public, who are entitled to demand: whose countryside is it anyway? Is our countryside a national asset for the whole world to enjoy? Or is it the exclusive preserve of a few very rich people who want to kill stuff?

To oppose the killing of birds of prey on grouse moors is regarded as a controversial and dangerous position. The controversial stance is that people should abide by the law. People who hold to this position in public can be guaranteed to be attacked by a very powerful lobby; I have some bruising personal experience here. What the lobby is saying, in effect, is that we have the right to break the law. The laws of the land don't apply to us. We can pick and choose the laws we keep. We can do what we like, because we're rich and you're not.

Anger seems to me a helpful response in these circumstances.

facing page:
A HERON,
(bottom right)
accompanied by
a bittern (left) and
an egret (top left).

We need to be concerned not only for the birds, not only for the wild world, but also for ourselves and for our children. We are becoming nature-deprived, and we are bringing up a generation of nature-deprived children. . . the first generation in history with a lower life expectancy than its parents. The *Oxford Junior Dictionary* has chucked out such words as adder, heron, kingfisher, minnow, thrush, blackberry, bluebell, bramble and poppy, and brought in blog, broadband, voicemail, attachment, database, chatroom, bullet point and cut-and-paste.

A generation ago, 40 per cent of children played regularly in natural areas; the current figure is 10 per cent. It's been established that 40 per cent of children never play outdoors at all. The problem of attention deficit hyperactivity disorder (ADHD) in children is a great worry to educationists and is often treated with drugs; outdoor activity appears to reduce the symptoms of ADHD by 30 per cent in urban surroundings and by 300 per cent in a rural setting.

In other words, nature-deprivation isn't just about acorns. It's about mental and physical health; spiritual health too, if you like. It's intimately associated with happiness. The wild world helps us to deal more easily with pain and loss, and allows us to take more pleasure and satisfaction in the lives we lead.

We need the wild world as much as the wild world needs us. We are mutually dependent. As we destroy the wild world, so we destroy ourselves.

It's a good thing, then, that there are people out there doing something about it.

right:
THE RHINOCEROS
HORNBILL.
One of the world's most
improbable creatures.

What if you were only allowed one bird? In some moods I'd take the rhinoceros hornbill—one of the world's most improbable creatures. I'd previously confused this with the pied hornbill, which has an impressive casque on the top of its head above a pretty enormous bill. Ha! That's nothing! When I saw a rhino hornbill I entered a different world of improbability. The bill was absurd, and the casque—up-tilted at the front to resemble a rhino horn—lay so far into the realms of fantasy that the human imagination can hardly cope with it.

And then it flew. A very big bird, on immensely broad and powerful wings, making loud, powerful swishes as it moved through the air—but how come it didn't nose-dive like a paper dart with a paperclip on its nose? How come it didn't land in the earth beak-first and stick there with its feet kicking helplessly at the air? Answer: because the bill and casque are super-light, and for all its outlandishness, this is a properly evolved and wholly viable animal. The casque, worn by both sexes, is important—it's how the birds recognize each other and know what species they belong to. I saw six species of hornbill along the Kinabatangan River: ring-tailed, black, rhinoceros, bushy-crested, Oriental pied and white-crowned. They could possibly be confused, certainly at a distance, but the differences of casque, bill and face markings make things quite clear. That matters even more to a hornbill than it does to a birder.

I saw this amazing bird while visiting a truly amazing organization. Hutan began in the jungles of Borneo as a research project, gathering information about orang-utans. But the directors of the project, Isabelle

Lackman and Marc Ancrenaz, soon saw that pure research was a luxury that the forest and the orang-utans couldn't afford.

So they established a community conservation movement, based at Kampung Sukau. The research continues, and many of the local people are involved. I met a group of them, all with clipboards in hand, recording the movements and behaviour of a mother orang and baby. These people are trained ethologists. When they attend international conferences on primatology, they are frequently asked where they got their doctorates. 'Sukau,' they reply modestly.

There are also people involved in forest restoration and tree planting. I met a group of wonderfully feisty ladies deep in the forest getting on with the job. It turns out that women are better at this than men, being more inclined to the task of nurturing.

There is also a task force that keeps elephants away from villages and plantations. Chilli fencing is effective here—the elephants don't like the pepper juice—and they also use a sort of drain-pipe cannon that makes a nice big bang and usually sends the elephants on their way, enabling humans and elephants to share the same space without squabbling. There is also a squad of forest rangers who patrol the paths and the rivers; their presence alone is a deterrent to poachers and illegal loggers.

The project acquired rights over a cave full of nesting cave swiftlets (their nests are used to make bird's nest soup). This site had been over-exploited and the population was dwindling; under the new management, the swiftlets were building up again. There is a plan to start harvesting the nests sustainably—and profitably—once the numbers have built back up again. Should you wish to visit, you can organize a trip into

I met a group of wonderfully feisty ladies deep in the forest getting on with the job

the forest to spend some time with orang-utans, in the same controlled way that you can visit gorillas. The profits go to the local community.

Some project. Good people doing good things in a good place. The world is full of them.

'Isabelle?'

'Yes?'

The boat was slipping along the Kinabatangan River and we were in a mellow mood, sated with wonders.

'What would you do if you had a million pounds?'

Isabelle laughed. 'I would buy an area of forest—I can show it to you—that would link up two areas of forest we already own. It is the best land, fronting on to the river, essential for wildlife. If we could secure that—well, it would be a dream.'

A year later, we gave her that million quid, and she bought the land.

The 'we' in question was the World Land Trust, a conservation organization based in Halesworth in Suffolk. I am a council member. The Trust's principal task is to finance land purchase on behalf of highly motivated cash-strapped conservation organizations in the developing world. It doesn't own an acre of land outside the UK, but it makes land purchase possible all over the world.

The money for Hutan was raised on appeal. Business organizations and private individuals joined together. I gave it plenty of space in the column I was then writing for *The Times* and the readers—thanks a million (literally) if you were one of them—came up trumps.

Great people doing great things in great places financed by a great organization far away and supported by great-hearted people who believed that saving a rainforest was a good thing. Good stuff, yes?

What a ridiculous bird. Gangly-legged and goggle-eyed—a bit like Alice in *The Vicar of Dibley*—it was standing in the middle of a parched and stony field with an air of quiet bewilderment. A wonderful thing entirely: a stone curlew—a night forager, hence the mad eyes, goggly for improved night vision.

The idea of seeing so enormous a bird in Britain seemed for a moment impossible: it was ridiculous, must have been a hallucination

They are one of the suites of birds that have adapted their behaviour to nest on farmland, as their favoured wild places were destroyed, and then found themselves in trouble when farming practices changed. Stone curlews are mostly confined to the Brecks of East Anglia and to Salisbury Plain.

They are making the beginnings of a comeback. In 2015 I spent the day, as mentioned in a previous chapter (see page 277), with Nigel Cowan, project manager of the East England Stone Curlew Recovery Project. He was looking for nests (a tricky business since the bird's entire strategy is based on keeping the nest hidden) so that the farmer would know where the nests were and could drive the tractor around them—rather than squishing the nests as has often happened in the past.

Many farmers have accepted subsidies from Defra to farm less intensively—to give the birds houseroom. I met one of them. Chris Cock said: 'It's nothing detrimental to the farm. It's about land management. And it's a bit more interesting than talking about sugar-beet prices.'

Conservation management is pretty intense right now, but there is an exit strategy. Farmers are being encouraged and are accepting subsidies to leave open patches on their land. I saw one such, with a stone curlew surely and confidently (if secretively) nesting right in the middle of it. Also the Stanford Training Area, owned by the Ministry of Defence, is now managing the land in a manner that suits stone curlews.

A good project in an arid landscape. The world is full of them.

I was on the Isle of Mull off the west coast of Scotland, on a boat in a sea loch. We waited long enough to savour that biting disappointment that hardened wildlifers learn to treat with breezy resignation. It wasn't going to show, was it? Damn and blast and all that, but that's (wild) life. And then all at once it was as if the entire sky was blotted out. The idea of seeing so enormous a bird in Britain seemed for a moment impossible: it was ridiculous, must have been a hallucination, there must be a rational explanation. Perhaps a giant had thrown a wardrobe from the top of the surrounding hills.

But no, it was a white-tailed eagle. The white-tailed eagle was shot to extinction in this country and the idea of it returning was the stuff of fantasy, along with ideas of bringing wolves back to the Highlands.

But it happened. A reintroduction programme for white-tailed eagles began in 1975 on the island of Rum and now the birds breed across the Western Isles and along the western coast. They have also brought millions to Mull; the eagles are part of the tourism industry, a major attraction for visitors. Very few of these visitors are birdwatchers in any demanding sense of the term. But they were all mad to watch an eagle or two, and I caught up with visitors at a forest hide and on two separate boat trips. A bird that was hated to death seems to have been loved back to life. A rum business, and a deeply cheering one.

A couple of years ago I was asked to make a nomination for the poll to find Britain's national bird. I suggested swallow: a bird that brings joy, a bird that is crucial to the British landscape, and yet as a long-haul migrant it's a bird we must share with the world. The swallow makes it clear that birds are a national and an international responsibility. It never made the shortlist, alas, though I still think it was the right choice.

But the public made their decision and here is the list they came up with, with the flyaway winner first.

1. Robin
2. Barn owl
3. Blackbird
4. Wren
5. Red kite
6. Kingfisher
7. Mute swan
8. Blue tit
9. Hen harrier
10. Puffin

Hen harrier made it so high as a result of some rather mischievous lobbying from the angrier side of the conservation groups, and that kept the bird in the area of public debate, which is a good thing. The red kite is

worthy of note, because it was shot to extinction as a breeding bird in England, but has made a comeback (see page 290), and is now seen as a national asset.

But as we celebrate the robin's inevitable victory, it's worth noting that 200,000 people bothered to vote, and that the stunt received national coverage. It seems that birds matter. It doesn't take very much to get birds in the news. We care about them.

above: THE BARN OWL was runner-up to the robin in a 2015 poll to find the UK's 'national bird', with 12 per cent of votes cast.

Among my Norfolk neighbours is the Raveningham Estate of Sir Nicholas Bacon, and every few months I have the pleasure of making a trip around their great acreage of land in the company of the estate manager, Jake Fiennes. The land is managed in a hard-headed fashion. It is highly successful in terms of agriculture and it is also managed for wildlife. The management of the grazing marshes has been so successful that it has produced more breeding waders than the nearby RSPB reserves... And it is here that my proposed survey route for the British Trust for Ornithology lies. No, next spring. Seriously.

The land management is done with the help of EU grants and is performed with genuine commitment. Here in private hands there is acre upon acre of land that is jumping with wildlife, most obviously with birds.

One of the birds you're almost certain to see on the Raveningham marshes is a marsh harrier. That's not so obvious as it sounds. A few miles away, I often see marsh harriers from the place where I work. It's astonishing to recall that this species was down to a single breeding pair in this country, back in 1971. A combination of good land management, a relaxation of illegal persecution and the reforms in the laws about pesticides have made this extraordinary comeback possible. It's now estimated that there are around 400 breeding pairs of marsh harriers. In the right place they are almost common. If you ever feel despair about the state of the world, an hour or so at an East Anglian marsh will cheer you up.

'I've just seen the biggest bloody herons I have ever seen in my life.'
In September 1979, a tenant farmer with cattle on part of the Horsey
marshes spoke those words to John Buxton, a wildlife film-maker who
owned the Horsey Estate in North Norfolk. These herons turned out to
be three cranes: birds who had got a little mislaid on their migration from
Scandinavia to their wintering grounds in southern France and Spain.

Since then they have stayed—and, at the speed of a glacier, they have

left:
THE CRANE
has made a comeback
in England, with a small
population of breeding
birds in Norfolk and
the Fens.

spread. There is now a tiny population of breeding cranes in England. They have spread across the Norfolk Broads and into the Fens. These vast birds, standing up to 1.6 metres (5.2 ft) in height and with a wingspan of 2.3 metres (7.5 ft), bigger than a white-tailed eagle, are now part of British life. You can meet them yourself if you pay a visit to the Norfolk Wildlife Trust's reserve at Hickling Broad.

Lakenheath Fen was a carrot field twenty years ago. It has been brilliantly restored to its original state—restored to itself—with such success that cranes flew in of their own accord and started breeding. Since then there has been a reintroduction project on the Somerset Levels.

Cranes were extinct in Britain for about 500 years, after playing a role in too many medieval banquets. But when conditions were right, they dropped back in again and spread. And with human help, they are spreading again.

Cranes were extinct in Britain for about 500 years, after playing a role in too many medieval banquets

Let's go back to the Fen Country. One of the last bits of fen we've got left, Woodwalton Fen, was bought by the great pioneer conservationist Charles Rothschild, whom we've already met (see page 242). A few miles away lies another isolated chunk of relict fen, Holme Fen. And someone thought: wouldn't it be a nice idea to join them up? To bring one vast sweep of ancient fen back to life?

And that's exactly what's happening. The Great Fen Project plans to create a wilder fenland landscape: an area of 3,700 hectares, or more than 14 square miles (36 sq km). Already the project has acquired 55 per cent of the land. It's a slow business, because there are sitting tenants. But the opposition has been overcome and the place is making a slow but relent-less march back to becoming itself again. The great draining of the fens destroyed more than 99 per cent of the mysterious watery habitat. Now we're putting some of it back.

This is a classic example of what's called landscape-scale conser-vation, and it's happening across the country. There are now literally hundreds of projects seeking to join up wild places and create a wilder, richer landscape across the country: Isle of Eigg, Coigach and Assynt, Anglesey Wetlands, Gwent Levels, Belfast Hills, River Don catchment, Lakeland, Lincolnshire Wolds, Oswestry Hills, Romney Marshes and North Somerset: just a few examples.

I had hoped to do a wilder and more exclusive trip, but the weather was against me. The sea was far too lumpy for a safe landing. So I got my-self rebooked on to a tourist trip—and that gave me a revelation of a quite different sort. I was in the town of North Berwick on the south shore of the Firth of Forth, and I was paying a visit to what is perhaps the single most dramatic wildlife spot in all of Britain.

Bass Rock is home to about 75,000 pairs of breeding gannets: vast birds with spear-tipped beaks and a wingspan of 1.9 metres (6 ft). You can see the rock rising sheer from the sea and standing 300 feet proud of the waves, like a vast wedding cake, white not with guano but with the birds themselves. You can see it from the middle of town. You can get a better view from one of the town's two golf courses, and a better one still if you take the boat.

You book your trip at the Scottish Seabird Centre, a well-set-up place where you can buy a baked potato and a soft-toy gannet. I went out on the *Seafari Explorer* with a jolly crowd of tourists, which included babes in arms and a tangerine-coloured spaniel. The trip was about an hour, and it took us to the foot of these towering cliffs. When we got there it was like being suddenly surrounded by angels, powering in the sky all around us on majestic snow-white wings—tipped with black, because, as attentive readers will recall, melanin gives strength to a feather, and you need added strength at the tips.

Only one other person on the boat had binoculars. The rest were just there because they wanted to see gannets. They wanted an hour with the angels. And together we looked and we gazed and we gasped. We saw gannets plunging into the sea from 30 metres (100 ft). We saw them quarrelling amiably and jostling for space on the crowded cliffs. We saw them launching themselves gloriously into the air. We saw their rather more awkward landings, for they are built for speed and they can find it tricky to stall out of the air with perfect precision in the gusty eddies around the cliffs.

There was a wonderful and elated sense of sharing. Here was one of the world's great wonders and we had all taken part in it. We were there, simply enough, because it was fab. Because it was joyous. Because birds matter. And as we turned back to town, a gannet kept pace with the boat for while, flying at exactly the same pace, as if he was indulging in a little human-spotting.

Birds need people these days. We can't duck that. They need us to make room for them, to look after their food supplies, to look after the places where they breed and where they rest and where they winter.

Birds need people these days. We can't duck that. They need us to make room for them, to look after the places where they breed

But throughout history, birds have completed us. Birds have told us about flight and colour and music and place and time. We have killed birds for food and for sport. We have domesticated birds and made them part of our lives and ourselves. Birds have helped us to understand the world by means of myth and symbol and story, and birds have taught us to understand the world through scientific thought and observation. Birds have taught us to understand about loss. Birds have shown us how we can restore the world to itself. And, above all, birds have given us joy.

We ought to look after birds because it's our clear moral duty to do so. But when we do this job we do it also for ourselves. Birds need people, yes, we know that.

But here's another fact. People need birds.

Birds mentioned in the text

right:
DARWIN'S
TREE OF LIFE.
'It shows how one species
can produce another,
or more than one other.
It also shows that some
lines will go extinct.
Above this [Darwin]
wrote two words in his
decisive scrawl.
"I think."'

(see page 201)

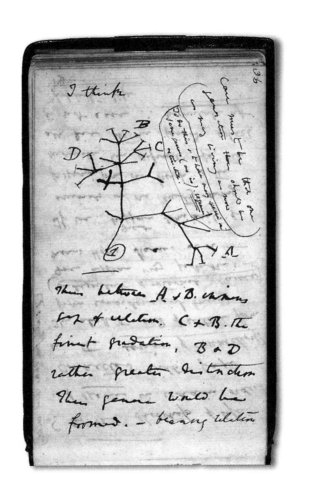